Competency Based Mathematics

for

Secondary Schools

Book 3

(MODULES 10 TO 14)

Nji Emmanuel Ndi
GBHS Mankon - Bamenda
North West Region Cameroon
Tel: (+237) 676 684 050
Email: manuelndike@gmail.com

First Edition

Printed by CreateSpace, an Amazon.com Company

EStore address: www.CreateSpace.com/7055960

Available from Amazon.com, CreateSpace.com, and other retail outlets

Available on Kindle and other retail outlets

Books by Nji Emmanuel Ndi

Complete Ordinary Level Mathematics Passport

Rudiments of Ordinary Level Mathematics

Advanced Level Pure Mathematics Key Facts

Competency Based Mathematics for Secondary Schools Book 1

Competency Based Mathematics for Secondary Schools Book 2

Competency Based Mathematics for Secondary Schools Book 3

Competency Based Mathematics for Secondary Schools Book 4

Competency Based Mathematics for Secondary Schools Book 5

Copyright © 2017 Nji Emmanuel Ndi

All rights reserved.

ISBN-10: 1545071640

ISBN-13: 978-1545071649

DEDICATION

Dedicated to all emerging and emergent Societies

Competency Based Mathematics for Secondary Schools. Book 3

Table of Contents

DEDICATION III
    Acknowledgement ix
    How to Use this Book x
    Notations Used in this Book xii

MODULE 10: 1

NUMBERS, FUNDAMENTAL OPERATIONS AND RELATIONSHIPS IN THE SETS OF NUMBERS AND BETWEEN ELEMENTS OF SETS 1

TOPIC 1 : SETS 2

1.1	THE SETS OF NUMBERS	3
1.2	THE NOTION OF SETS	3
1.3	SET NOTATION AND DEFINITION OF SETS	4
1.4	THE CARDINALITY OF A FINITE SET	7
1.5	TYPES OF SETS	7
1.6	THE UNIVERSAL SET	9
1.7	SUBSETS	9
1.8	EQUIVALENT SETS AND EQUAL SETS	11
1.9	INTERSECTION AND UNION OF SETS	12
1.10	DISJOINT SETS	13
1.11	REPRESENTATION OF SETS – VENN DIAGRAMS	13
1.12	THE COMPLEMENT OF A SET $C_\mathscr{E}^A$ OR A'	15
1.13	THE POWER OR DERIVED SET OF A SET	16
1.14	ALGEBRAIC LAWS OF SETS	18
1.15	CARDINALITY LOGIC	18
1.16	TRANSCRIBING FROM SET LANGUAGE TO ORDINARY LANGUAGE	22

TOPIC 2 : INDICES AND LOGARITHMS 27

2.1	MEANING OF b^p	28
2.2	LAW OF INDICES	28
2.3	EXPONENTIAL OR INDEX EQUATIONS	31
2.4	DEFINITION OF LOGARITHMS	32
2.5	LOGARITHMIC EQUATIONS	33
2.6	LAWS OF LOGARITHMS	34

TOPIC 3 : MATRICES 39

3.1	CONCEPT AND REPRESENTATION OF MATRICES	40
3.2	DEFINITION AND SIZE OF A MATRIX	40
3.3	TYPES OF MATRICES	41
3.4	EQUIVALENT MATRICES AND EQUAL MATRICES	43
3.5	ADDITION AND SUBTRACTION OF MATRICES	45
3.6	SCALAR MULTIPLICATION OF MATRICES	46
3.7	MULTIPLICATION OF MATRICES	47

Table of Contents

3.8	THE TRANSPOSE OF A MATRIX	50
3.9	THE DETERMINANT OF A 2×2 MATRIX	51
3.10	SINGULAR MATRICES	52

MODULE 11: **55**

PLANE GEOMETRY **55**

TOPIC 4 : CONGRUENCY AND SIMILARITY **56**

4.1	CONGRUENT FIGURES	57
4.2	CONGRUENT TRIANGLES	57
4.3	SIMILAR FIGURES	60
4.4	SIMILAR TRIANGLES	60
4.5	RATIO OF AREAS OF SIMILAR FIGURES	63
4.6	RATIO OF VOLUMES OF SIMILAR FIGURES	66

TOPIC 5 : VECTORS **76**

5.1	VECTOR AND SCALAR QUANTITIES	77
5.2	VECTORS AS DIRECTED LINE SEGMENTS	78
5.3	VECTOR NOTATION	78
5.4	COLUMN VECTORS	79
5.5	THE MAGNITUDE OR MODULUS OF A VECTOR	80
5.6	EQUALITY OF VECTORS	82
5.7	FIXED AND FREE VECTORS	83
5.8	ADDITION OF VECTORS	86
5.9	SUBTRACTION OF VECTORS	88
5.10	MULTIPLICATION OF VECTORS BY SCALARS	92

TOPIC 6 : TRIGONOMETRY **96**

6.1	MEANING OF TRIGONOMETRY	97
6.2	STANDARD NOTATION FOR TRIANGLES	97
6.3	THE RIGHT-ANGLED TRIANGLE	97
6.4	THE PYTHAGORAS THEOREM	98
6.5	TRIGONOMETRIC RATIOS	101
6.6	TRIGONOMETRIC RATIOS FROM CALCULATORS	104
6.7	ACUTE ANGLE TRIGONOMETRIC RATIOS	107
6.8	TRIGONOMETRIC RATIOS FROM TABLES	108
6.9	INVERSE TRIGONOMETRIC RATIOS	109
6.10	FINDING OTHER TRIG RATIOS GIVEN ANOTHER	110
6.11	COMPLEMENTARY ANGLES	112
6.12	TRIG RATIOS OF SPECIAL ANGLES	114
6.13	REAL LIFE APPLICATION OF TRIGONOMETRY	117

MODULE 12: **124**

SOLID FIGURES **124**

TOPIC 7 : MENSURATION OF SOLIDS **125**

7.1	Surface Area and Volume of Prisms	126
7.2	Cones and Pyramids	127
7.3	Surface Area and Volume of a Sphere	131
7.4	Composite Solid Figures	132
7.5	Volume and Surface Area of a Frustum	132
7.6	Volume and Surface Area of Hemisphere	135

MODULE 13: STATISTICS AND PROBABILITY 141

TOPIC 8 : STATISTICS 142

8.1	Frequency-Distribution Tables	143
8.2	Representation of Data-Statistical Graphs	143
8.3	Measures of Central Tendencies	148
8.4	Arithmetic Mean (Average or Mean)	149
8.5	Median	151
8.6	Choicest Measure of Central Tendency	153
8.7	Grouped Data	155
8.8	Histograms for Grouped Data	156
8.9	Finding the Mode by Calculation	159
8.10	Frequency Distribution Curve	159
	(Frequency Polygon)	159

TOPIC 9 : PROBABILITY 172

9.1	The Concept of Probability	173
9.2	Some Basic Probability Terminology	173
9.3	Probability as a Number	175
9.4	Equiprobable Outcomes	175
9.5	Standard Definition of Probability	176

MODULE 14: 180

ALGEBRA AND LOGIC 180

TOPIC 10 : SIMPLE ALGEBRA 181

10.1	Review of Algebraic Expressions	182
10.2	Expansion of a Product of Two Binomials	182
10.3	Expansion of the Square of a Binomial	183
10.4	Simple Factorization	184
10.5	Factorization by Grouping	185
10.6	The Difference of Two Squares	185
10.7	Factorizing Quadratic Expressions	187

TOPIC 11 : EQUATIONS 192

11.1	Simple Linear Equations	193
11.2	Simultaneous Linear Equations	193
11.3	Simultaneous Equations with Uniform Coefficients	194
11.4	Simultaneous Equations with Non-uniform Coefficients	198

11.5	Simultaneous Equations with Fractions and Decimals	200
11.6	Simultaneous Linear Equations in Real Life	203
11.7	Quadratic Equations	206
11.8	Standard Form Quadratic Equations	206
11.9	Factorization method	207
11.10	The Quadratic Formula	208
11.11	Problems leading to Quadratic Equations	209

TOPIC 12 : LOGIC 215

12.1	The Concept of Logic	216
12.2	Statements or Propositions	216
12.3	The Truth Value of a Statement	216
12.4	Closed and Open Statements	217
12.5	Domain and Variable	217
12.6	Negation, $\sim p$	219
12.7	Compound or Composite Statements	220
12.8	Conjunction	221
12.9	Disjunction	221
12.10	Conditional Statements, $p \to q$	223
12.11	Biconditional Statement, $p \Leftrightarrow q$ or $p \leftrightarrow q$	224
12.12	Logical Equivalence	225
12.13	De Morgan's Laws	227
12.14	Connectors	227
12.15	Tautologies	229
12.16	Contradictions	229
12.17	Quantifiers	229
12.18	Syllogisms	232
12.19	Hypotheses and Conclusions	232

TOPIC 13 : TRANSPOSITION OF FORMULAE 237

13.1	Making a Subject of a Formula	238
13.2	Formulae without Square Roots	238
13.3	Formulae Containing Square Roots	240
13.4	Formulae Containing Quadratics	242

TOPIC 14 : VARIATION 246

14.1	Real Life Examples of Variation	247
14.2	Direct variation	247
14.3	Inverse variation	251
14.4	Other Inverse Variations	253
	Again, the y-axis and the x-axis are asymptotes to the curve.	254
14.5	Joint or Combined Variation	255

TOPIC 15 : RELATIONS AND FUCTIONS 261

RELATIONS		262
15.1	The Idea of a Relation	262

15.2	The Cartesian product of Two Sets	262
15.3	Mathematical Relation	263
15.4	Notation	264
15.5	Ways of Defining Relations	264
15.6	Inverse Relation	266
15.7	Properties of Relations in a Set	270
15.8	Equivalence relation	272
15.9	Order Relation	273

MAPPINGS AND FUNCTIONS 273

15.10	The Idea of a Function or a Mapping	273
15.11	Function Notation	274
15.12	Representation of Functions	274
15.13	Types of Mapping	278
15.14	Flow Charts	283
15.15	Inverse Function	283
15.16	Composite Functions	284
15.17	Restricted Domain and Restricted Function	285

ANSWERS TO STRUCTURAL EXERCISES **295**

Acknowledgement

My deepest gratitude goes to God Almighty for the inspiration and for the strength.

Many thanks go to Mme. Mbuameh Daisy and Mr. Mburubah Walters for their critical proof reading of the typescript and for offering very useful suggestions which went a long way to reshape the work, the North West Regional Pedagogic Inspector for Mathematics Mr. Nfor Samuel Ndi who preview the initial manuscript and gave ample advice, which went a long way to reshape the document. I heartily thank the Former North West Regional Pedagogic Inspector for Mathematics Mr. Nji Samuel Tatah who made a very commendable effort to edit the Mathematics content of the book. I cannot forget the last minute encouragements and advice which the National inspector of Mathematics Mme Babila Emilia inspired me with. I equally pay much tribute to my students on which this material was tested. I cannot end here without thanking my sweet heart Nji Irene Nfih and my Children who encouraged and supported me in one way or the other during the course of the work.

Many thanks go to the WAEC and the CGCE Board for allowing their past questions to be used directly or indirectly.

Nji Emmanuel Ndi
G.B.H.S. Mankon, Bamenda
North West Region
Cameroon
TEL: (+237)76684050
E-mail: manuelndike@gmail.com

Competency Based Mathematics for Secondary Schools. Book 3

How to Use this Book

This book is written in a very special way with different sections boxed and represented by special symbols as follows.

| ? | **Brainstorming Exercise** |

| ✎ | **Example** |

| 👍 | **Real life Examples** |

| 🎬 | **Exercise** |

| 🛠 | **Skill Building Exercise** |

| 💬 | **Discussion Exercise** 💬 |

| 🏭 | **Integration Activity** |

| 🔍 | **Investigative Activity** |

| ✍ | **Multiple Choice Exercise** |

| 📄 | **Review Exercise** |

How to Use this Book

 Group Activity

The various sections represented by different symbols are out to facilitate navigation through the book. By investing enough time and energy in each section both students and teachers will realize that their speed and understanding will be greatly enhanced.

The brain storming exercises are aimed at provoking and invoking the learners' minds to prepare them for the task at hand. The teacher is highly encouraged to orally question the students during lessons using questions under this section.

The investigative exercises are meant to give the learner ample opportunity to experiment and self discover facts and concepts and develop methods and skills without being told.

The group activities and discussion exercises are aimed at developing a team spirit in the learners.

Many well designed examples are vividly used and solved to facilitate the learner's understanding by showing the necessary steps required for a particular solution. There are a good number of real life examples which point out the application of the subject matter in real life situations. The student is advised to study these examples very carefully.

There are many well graded exercises and skill building exercises to test the level of understanding of the learner and to facilitate skill development in the learner. The student is advised to attempt all the questions as each question may have its own technique.

Many integration activities have been designed to unify groups of sub topics, topics or modules in some cases.

Where necessary review exercises have been given to help the learner retain the skills acquired in the earlier sections.

Finally each topic ends with a good number of multiple choice questions. In each question only one of the alternatives is correct. Write down the letter corresponding to the correct answer.

For greatest achievement, the learner is advised to study regularly what he does not know and work without fear of making mistakes whether with the teacher or during group work.

By consistently and systematically going through this course as instructed, the learner will be overwhelmed with the competencies acquired at each level and at the end of the course.

Competency Based Mathematics for Secondary Schools. Book 3

Notations Used in this Book

Notation	Meaning
$\{\ldots\}$	The set of elements…or the unordered list with elements…
$n(A)$	The number of elements in set A
$\{x:\ \}$	The set of all x such that
$\in$	Is an element of …
$\notin$	Is not an element of …
$\{\ \}$ or $\emptyset$	The empty set.
$\mathscr{E}$	The universal set.
$\cup$	The union of …
$\cap$	The intersection of…
$\subseteq$	Is a subset of …
$\subset$	Is a proper subset of …
$A \backslash B$	The difference between the sets A and B.
$(a, b, c, \ldots)$	An ordered list of elements a, b, c, …
$\{a, b, c, \ldots\}$	The set or an unordered list of elements a, b, c, …
$\mathbb{Z}$	The set of integers, $\{0, \pm 1, \pm 2, \pm 3, \pm 4, \ldots\}$
$\mathbb{N}$	The set of all positive integers and zero, $\{0, 1, 2, 3, 4, \ldots\}$
$\mathbb{Z}^+$	The set of positive integers $\{+1, +2, +3, +4 \ldots\}$
$\mathbb{Q}$	The set of rational numbers
$\mathbb{Q}^+$	The set of positive rational numbers
$\mathbb{R}$	The set of all real numbers $\{x: x \in \mathbb{R}\}$
$\mathbb{R}^+$	The set of all positive real numbers $\{x \in \mathbb{R}: x > 0\}$
$f(x)$	f of x or the image of x under the function f
f^{-1}	The inverse function of the function f
fg or $f \circ g$	The function f of the function f
$=$	Is equal to
$\neq$	Is not equal to
$\approx$	Is approximately equal to
$<$	Is less than
$>$	Is greater than
$\not<$	Is not less than
$\not>$	Is not greater than

$\leq$	Is less than or equal to
$\geq$	Is greater than or equal to
$a < x < b$ or $]a, b[$ or (a, b)	An open interval on the number line
$a \leq x \leq b$ or $[a, b]$	A closed interval on the number line
$\{x : a < x < b\}$	The set of elements x such that a is less than x and x is less than b
a	The vector **a**
AB	The vector represented in magnitude and direction by AB
$\|x\|$	The modulus or absolute value of x i.e. $\{x \text{ for } x > 0, -x \text{ for } x < 0, x \in \mathbb{R}\}$
$\boldsymbol{a} \cdot \boldsymbol{b}$	The dot or scalar product of the vectors **a** and **b**
A^{-1}	The inverse of the non-singular matrix A
A^T	The transpose of the matrix A
lg x or log x	The common logarithm of x
x^n	The number x, raised to the power n
$\propto$	Is proportional
∞	Infinity
$\sqrt{}$	The positive square root
$-\sqrt{}$	The negative square root
$\sqrt[n]{a}$	The nth root of a
p:	The statement or preposition p
T or 1 in truth tables	True
F or 0 in truth tables	False
$\sim p$ or p' or $\neg p$	The negation of a statement p
$p \wedge q$	The conjunction of the statements p and q
$p \wedge q$	The disjunction of the statements p and q
$A \cap B$ or $\{x : x \in A \wedge x \in B\}$	The intersection of sets A and B.
$A \cup B$ or $\{x : x \in A \vee x \in B\}$	The union of sets A and B.
$p \Rightarrow q$ or $p \rightarrow q$	p implies q or p is sufficient for q or p only if q or q is necessary for p
$p \Leftrightarrow q$ or $p \leftrightarrow q$ or p iff q	p is a necessary and sufficient condition for q or p implies and is implied by q or p if and only if q

$\forall x$	For all or for every element x
$\exists x$	There exists or for at least one or for some element x
$\exists! x$	There exists one and only one element x
$\equiv$	Is equivalent or is congruent to
///	Is similar to
$\perp$	Is perpendicular to
$\parallel$	Is parallel to
$G = (V, E)$	The graph of the set V of vertices together with the set E of edges
$D = (V, A)$	The directed graph (digraph) of the set V of vertices and the set A of ordered edges
$G = (V, E, A)$	The mixed graph of the set of vertices V, unordered edges E and ordered edges A.
$n :=$	The store n takes the value…
°	Degree
°C	Degrees Celsius
°F	Degrees Fahrenheit

Module 10

Numbers, Fundamental Operations and Relationships in the Sets of Numbers and between Elements of Sets

Family of Situations
Module 10 is an extension of modules 1 and 5. At the end of this module; the student is expected to have acquired more competencies within the **families of situations** *'Representation, determination of quantities and identification of objects by numbers'*.

Categories of Action
The categories of action for module 10 include:
1. Determination of a number,
2. Reading and writing information using numbers,
3. Verbal interaction on information containing numbers,
4. Representation and treatment of information and quantities.

Credit
The module is expected to be covered within 5 weeks teaching 4 periods of 50 minutes per week (or within 20 periods of 50 minutes).

Topic 1

SETS

Objectives
At the end of this topic, the learner should be able to:

1. Define and identify the sets $\mathbb{N}, \mathbb{Z}, \mathbb{Q}, \mathbb{R}$.
2. Carry operations in each of the sets $\mathbb{N}, \mathbb{Z}, \mathbb{Q}, \mathbb{R}$.
3. Use set notations correctly.
4. Transcribe set language to ordinary language and vice versa.
5. Solve real life problems involving set theory.
6. Find the cardinality of a set.
7. Identify and differentiate between equal and equivalent sets.
8. State the compliments of a given set.
9. Find the number of subsets for finite sets.
10. Find the power set of finite sets with not more than 3 elements.
11. Draw and use Venn Diagrams.

1.1 The Sets of Numbers

 Review Exercise

1. Write down the symbol which denotes each of the following sets.
   (a) The set of natural numbers.   (b) The set of integers.
   (c) The set of rational numbers.   (d) The set of real numbers.
2. Describe the elements of each set and list 10 of them.
3. State the order of the operations $+, -, \times, \div$ in the set of real numbers.

Recall that:
1. The set of natural numbers is denoted by $\mathbb{N} = \{0, 1, 2, 3, 4, 5 \ldots\}$.
2. The set of integers is denoted by $\mathbb{Z} = \{0, \pm1, \pm2, \pm3, \ldots\}$.
3. A **rational number** is a number which can be expressed as a quotient of two integers. The set of rational numbers is denoted by $\mathbb{Q}$. Examples of rational numbers are $\frac{3}{4}, -\frac{5}{8}, \frac{7}{2}, 1, 0, -6, 1\frac{3}{8}$.
4. Irrational numbers are numbers that cannot be expressed as the quotient or ratio of two integers. We denote irrational numbers by $\mathbb{Q}'$.
5. The set of real numbers denoted by $\mathbb{R}$ is the set consisting of all the rational numbers and all the irrational numbers put together.
6. In the set of real numbers, the order of the basic operations $+, -, \times, \div$ is $\div, \times, +, -$ or **DMAS**.

1.2 The Notion of Sets

 Brainstorming Exercise

Let A = Monday, Tuesday, Wednesday, Thursday, Friday, Saturday, Sunday,
and B = blue, green, yellow, red, white, grey, black, violet, orange.

1. Describe each list without listing and explain what is common with the items of each list.
2. Given an item, is it possible to say whether or not the item qualifies to belong to any of the lists?
3. Is January qualified to be a member of list A? Why?
4. Is blackboard qualified to be a member of list B? Why?
5. What general principles are required to compose such lists?

Clearly, the first list A consist of days of the week and the second list B consist of colours. Also January is not a day of the week, so it cannot be included in the list A. Equally a blackboard is not a colour so it cannot be included in the list B. The above lists are examples of **sets**. We usually enclose such a list, which represents a set in braces, as below

A = {Monday, Tuesday, Wednesday, Thursday, Friday, Saturday, Sunday}.
B = {blue, green, yellow, red, white, grey, black, violet, orange}.

Therefore, a **set** is *a group of objects or data with common properties which distinguish them from every other entity.*

The objects or data that make up the set are called the **members** or **elements** of the set.

1.3 Set Notation and Definition of Sets

Capital letters A, B, C... are used to denote sets.
There are three basic methods for describing or defining sets.

The Roster Method

This is done by listing the elements of a set, separating them by commas and enclosing them with braces.

 Example

1. Let V = English vowels. List the elements of V.

   Solution
   $V = \{a, e, i, o, u\}$.

   If the list is so long, the first few elements are listed and three dots are used to show that the list continues.

2. Let E = positive even numbers (i.e. positive numbers that we can divide exactly by 2). List the elements of E.

   Solution
   $E = \{2, 4, 6, 8, 10...\}$

The Rule Definition Method

In this case, a rule, which qualifies any element as a member of the set, is stated or defined.

Module 10, Topic 1: Sets

 Example

Let C = {North, West, South, East}
R = {Red, Orange, Yellow, Green, Blue, Indigo, Violet}
Use the rule definition method to define C and R.

Solution
C = cardinal points
R = Colours of the rainbow.

 Brainstorming Exercise

Let X = Colours of the rainbow and Y = {Colours of the rainbow}.
List the elements of X and Y and explain the major difference between X and Y.

In using the rule definition method, it is unwise to use braces.

Colours of the rainbow ≠ {Colours of the rainbow}

The explanation is that X = {Red, Orange, Yellow, Green, Blue, Indigo, Violet} but Y is a set containing only one element. Its sole element is the phrase "Colours of the rainbow". To make this point clearer, consider the set.

P = {colours of the rainbow, days of the week}.

Clearly, P is a set of two elements. Its two elements are the phrases "colours of the rainbow" and "days of the week". A comma separates these two elements.

Set Builder Notation Method

 Example

1. Given that W is a set of days of the week. Use set builder notation to write down this statement.

   Solution
   $W = \{x : x \text{ is a day of the week}\}$
   or $W = \{x / x \text{ is a day of the week}\}$
   This is read, "W is the set of all elements, x, such that x is a day of the week".

   $x:x$ or x/x is read, 'x such that x'

2. Write in full the meaning of $P = \{(x, y) : y = 2x + 5\}$.

5

Solution

P is the set of all points (x, y) which satisfy the line $y = 2x + 5$.

3. State the meaning of $A = \{x: 0 \leq x \leq 3\}$.

   Solution

   A is the set of all points which lie between the points 0 and 3 inclusively.

Membership Notation

 Brainstorming Exercise

Let $V = \{x \,/\, x \text{ is an English vowel}\}$.
1. List the elements of the set V.
2. Make three different statements using different words (which are synonyms) which tell us that "a" is one of the items in the list.
3. Use symbols only, to rewrite the statement in 2.
4. Use symbols only, to write the opposite of the statement in 3.
5. Read your statement in 4; in three different ways.

Clearly, a belongs to the set V, or in other words, a is a member of V or a is an element of the set V. This fact is denoted by $a \in V$ and is read as "a belongs to V" or "a is an element of V" or "a is a member of V"

In mathematics, cancelling a sign changes its meaning to the opposite one. For instance, $\neq$ means, "is not equal to". Thus, we write 'b is not an element of V' symbolically as $b \notin V$, read, "b is not an element of V" or "b does not belong to V".

 Exercise 1:1

1. Identify the odd element in the following list of objects:
   (a) boy, girl, man, pen, woman
   (b) rice, ink, beans, maize, plantain
   (c) Cameroon, Nigeria, London, Egypt, France
   (d) h, j, i, p, q, w
   (e) Black, orange, yellow, white paper, green
2. Write the following statements using set notation:
   (a) x belongs to A (b) y is not a member of B (c) G has 3 members
3. Given that $F = \{a, b\}$. State which of the following statements is correct or incorrect giving reasons for your answer. (a) $b \in F$ (b) $\{b\} \in F$
4. Which of the following collection of objects form a set? Say why or why not.
   A = Months of the year with 30 days.

Module 10, Topic 1: Sets

B = Beautiful girls in form one.
X = Form three book list.
G = Good people in our school.
P = Cripples in our school.
I = Intelligent students in form two.
C = Black, yellow, white paper, green.

5. Given that $X = \{x: 3x = 12\}$ and $y = 4$. Is y equal to X? Give a reason for your answer.
6. Write the following sets using the roster method
   M = multiples of 2 less than 20
   F = suits of playing cards
   $V = \{x : x \text{ is a vowel of the English alphabet}\}$
   $A = \{y : y \text{ is a factor of } 36\}$
   $N = \{n \mid n \text{ is an integer between 1 and 10}\}$
7. Define a rule to qualify the members of each of the following sets.
   $A = \{1,3,5,7,9,11,13\}$
   $B = \{2,4,6,8,10,12,14,16,18,20\}$
   $C = \{\text{Mangoes, oranges, pineapples, bananas,...}\}$
   $D = \{\text{tall, great, intelligent, small,...}\}$
   $E = \{\text{football, table tennis, volleyball, athletics...}\}$
8. Write the following using set builder notation.
   $X = \{2,3,5,7, 11,13,17,19\}$
   $M = \{1,2,3,6,7,14,21,42\}$
   $C = \{\text{Momo, Mezam, Menchum, Boyo, Bui, Donga/Mantung, Ngohkentungia}\}$
   $Y = \{\text{Manyu, Sanaga, Ndian, Wouri, Mungo, Shari, Munaya, Mbam, Katsina}\}$
   $T = \{3,6,9, 12,15,18,21,24,27,30\}$

1.4 The Cardinality of a Finite Set

The cardinality or the cardinal number of a set or the order of a set is the number of elements in a set. For example, the cardinality of $V = \{a, e, i, o, u\}$ is 5.
This is denoted algebraically by $n(V) = 5$

1.5 Types of Sets

? Brainstorming Exercise

Let $X = \{4\}$, $Y = \{3,4\}$, $Z = \{2,3,5\}$, M = months of the year.
E = Students of form three who are four years old, W = whole numbers.

1. Write down the cardinality of each of the sets in the form $n(P) = q$.
2. Make statements in ordinary English which tell the cardinality of each of the set above.

> 3. Use the cardinality of each of the sets to deduce a name for each of the sets above.

Singleton or unit set

A **singleton** or unit set is a set that has only one element such as X above.

Doubleton or Pair set

A **doubleton** or a pair set is a set such as Y above whose cardinality is two.

A Trebleton

A **trebleton** is a set such as Z above whose cardinality is three.

Empty or null set

An empty or null set is a set such as E above which has no elements. We usually denote an empty set by the symbol $\emptyset$ or $\{\ \}$. Thus, $E = \emptyset = \{\ \} \Rightarrow n(E) = 0$.

Note that it is wrong to write $\emptyset = \{0\}$, $\emptyset = \{\emptyset\}$ or $\emptyset = \{...\}$ due to the following reasons. $\{\emptyset\}$ and $\{0\}$ are singletons, containing the elements $\emptyset$ and 0 respectively while, $\{...\}$ is a set whose elements have been omitted, or voluntarily left out.

Finite sets

A **finite set** is a set such as M above which has a fixed countable number of elements.

 Example

Let F = factors of 30. Explain whether F is a finite set.

Solution

$F = \{1, 2, 3, 5, 6, 10, 15, 30\}$ and $n(F) = 8$
Therefore, F is a finite set because it has a fixed countable number of elements.

Infinite sets

An infinite set is a set (such as W) with an inexhaustible or unlimited number of elements. The cardinality of an infinite set is infinity, meaning "extremely large". We denote infinity by the symbol ∞.

 Example

Let P = all the people since creation
And $E = \{x : x \text{ is an even number}\}$

What property has *P* and *E* in common?

Solution
Both *P* and *E* are infinite sets because they each contain an inexhaustible number of elements.

 Exercise 1:2

1. Classify the following sets as singleton, doubleton, trebleton, empty set or none of the above.
   (a) {1,3,5}   (b) {1}   (c) {0}   (d) {3,8}   (e) {0, ∅}
   (f) {0, ∅, 1}   (g) {2, 4}   (h) {a, e, i, o, u}   (i) {...}   (j) { }   (k) ∅
2. State the cardinality of each of the following sets.
   (a) {1,3,5,7,9,11,13}   (b) {3,6}   (c) { }   (d) {Students of your class}
   (e) {0}   (f) S = students of your class   (g) Factors of 42
   (h) Multiples of 5 less than 30   (i) Even numbers less than 30
   (j) Odd numbers less than 40
3. State whether the set is finite or infinite.
   (a) The set of whole numbers.   (b) ∅   (c) {0}
   (d) The set of even numbers   (e) The letters of the English alphabet
   (f) The set of odd numbers   (g) {x/x is a prime number less than 1000}
   (h) {$x:x$ is a planet}   (i) The set of teachers in your school
   (j) {1,2,4,8,16,32,64,...}
4. Fill in the blank spaces with ∈ or ∉
   (a) 2___{1,3,5,7,9,11,...}   (b) 14___{2,4,6,8,10,...}
   (c) 30___{$x:x$ is a multiple of 6}   (d) 17___{1,2,4,7,11}
   (e) Chalk___{paper, pencil, pen, ruler}
   (f) Beautiful___{go, take, come, eat, write}

1.6   The Universal Set

A **universal set** is a set which contains all the elements under consideration in a given circumstance or situation. We usually denote the universal set by $\mathscr{E}$ or *U*. In a given circumstance or situation all the sets under consideration contain only elements which are found in the universal set $\mathscr{E}$ or *U*.

1.7   Subsets

A subset is a set whose members are elements of another set. A **proper subset** of a set is a non-empty set whose members are elements of a bigger set called the **super set**. The empty set and the set itself are called the **improper subsets** of the set. If all the elements of a set A are found in another set B, then we say A is a subset of B, denoted by $A \subset B$ or $B \supset A$. A is not a subset of B is written $A \not\subset B$.

Example

1. Let $A = \{1,2,3,4,5,6,7,8\}$ and $B = \{2,5,7\}$.
   Write three statements relating A and B using the symbols
   (a) $\subset$ (b) $\supset$ (c) $\not\subset$.
   Which of these statements are the same?

   Solution
   (a) $B \subset A$ (b) $A \supset B$ (c) $A \not\subset B$
   The statement in (a) is the same as statement (b).

2. $\mathscr{E} = \{1,2,3,4,5,6,7,8\}$
   $A = \{2,4,6,8\}$
   $B = \{1,3,5,7,9\}$
   and $C = \{1,2,3,4\}$
   Which of the sets A, B and C is or is not a subset of $\mathscr{E}$? Explain.

   Solution
   The sets A and C are both subsets of the universal set $\mathscr{E}$ because all their elements in $\mathscr{E}$, while the set B is not because $9 \not\in \mathscr{E}$.

3. Given that, $A =$ Multiples of 3. List the elements of A when $\mathscr{E} =$ Whole numbers less than 30

   Solution
   $\mathscr{E} = \{1,2,3,4,5,6,7,89,10...,29\}$
   $A = \{3,6,9,12,15,18,21,24,27\}$

4. Given that,
   $\mathscr{E} = \{x : x \text{ is an odd number less than 30}\}$
   $C = \{x : x \text{ is a multiple of 3}\}$
       List the elements of set C.

   Solution
   $\mathscr{E} = \{1,3,5,7,9,11,13,15,17,19,21,23,25,27,29\}$
   $C = \{3,9,15,21,27\}$
   This is so since 6, 12, 18 and 24 are not odd numbers, hence not elements of C.

5. Let $X = \{1, 3, 5, 7, 9\}$
   $Y = \{2, 4, 6, 8, 10\}$
   $Z = \{1, 2, 3, 4, 5, 6, 7, 8, 9, 10\}$

   (a) State the proper subsets of Z.

(b) State the improper subsets of Z.
(c) List all the subsets of Z.

Solution
(a) The proper subsets of Z are X and Y.
(b) The improper subsets of Z are ∅ and Z.
(c) The subsets of Z are ∅, X, Y and Z.

1.8 Equivalent Sets and Equal Sets

Equivalent sets are sets which contain an equal number of elements.
If two non-empty sets X and Y are equivalent it means that $n(X) = n(Y)$ and we write $X \sim Y$.

Equal sets are sets which contain exactly the same number of elements and exactly the same elements. Thus if two sets X and Y are equal then $n(X) = n(Y)$ and $X \subset Y$ and $Y \subset X$. If two sets X and Y are equal, this is written algebraically as $X = Y$.

All equal sets are equivalent sets but equivalent sets are not necessarily equal sets.

Note here that *repeating elements of a set does not change the set.*

Thus, {2, 4, 6, 8} = {8, 2, 6, 4, 6, 8}

Example

If $X = \{3,6,9,12\}$ and $Y = \{2,4,6,8\}$. Say, giving reasons whether X and Y are equivalent sets.

Solution
$n(X) = n(Y) = 4$. Therefore X and Y are equivalent sets.

Exercise 1:3

1. Complete the following statements using the symbols ⊂, ⊄, =, ≠, ∼, ∈ or ∉.
   (a) $\{a,e,i,o,u\}$ ___ $\{x: x$ is a letter of the alphabet$\}$ (b) e ___ $\{a,e,i,o,u\}$
   (c) 3 ___ $\{x: x$ is an even number$\}$ (d) $\{3,7,2,5,1,6,4\}$ ___ $\{1,2,3,4,5,6,7\}$
   (e) $\{2, 4, 6\}$ ___ $\{1, 3, 5, 7, 9\}$ (f) $\{a, b, c\}$ ___ $\{1,2,3\}$

2. Let $\mathscr{E}$ = universal set and A, B, C, be any other sets such that
   $\mathscr{E} = \{1,2,3,\ldots,49,50\}$

A = multiples of 5
B = {1,2,3,4,5}
C = {1,2,3}
D = {3,5,1,4,2}
E = {3,4,5}
State whether each of the following is true or false.
(a) $A \subset E$ (b) $B \subset E$ (c) $B \in E$ (d) $C \not\subset E$ (e) $B \not\subset D$
(f) $E \subset B$ (g) $C \not\subset D$ (h) $55 \in A$ (i) $17 \notin E$
(j) $B \supset C$ (k) $B = D$ (m) $A \not\subset E$

3. List all the subsets of the set $P = \{1, 2, 3\}$. How many subsets, has P?
4. Which of the following sets are equal?
   $\{a, b, c\}, \{c, a, b, c\}, \{b, a, b, c\}, \{a, c, b, c\}$
5. Using only = or ~, write statements between the sets where applicable.
   (a) $\{1, 2, 3, 4, 5\}$ (b) $\{a, b, c, d, e\}$ (c) $\{2, 4, 6\}$
   (d) $\{4, 2, 5, 1, 3\}$ (e) $\{Biology, Chemistry, Mathematics\}$
6. How many proper subsets, has an empty set?
7. Given that $A = \{4\}$, $B = \{3,4\}$, $C = \{1,2,3\}$, $D = \{1,2\}$ and $E = \{1,2,4\}$. State whether each of the following statements is true or false.
   (a) $D \subset B$ (b) $B \neq E$ (c) $A \subset C$ (d) $B = D$
   (e) $B = C$ (f) $B \subset C$ (g) $B \sim D$ (h) $C \sim E$
8. Given that (i) $A \subset B, B \subset C$ (ii) $a \in A, b \in B$ and $c \in C$
   (iii) $d \notin A, e \notin B$ and $f \notin C$
   Which of the following statements must be true?
   (a) $a \in C$ (b) $b \in A$ (c) $c \notin A$
   (d) $d \in B$ (e) $e \notin A$ (f) $f \notin A$
9. State which of the following sets are equivalent or equal?
   (a) $A = \{a,b,c,d,e\}$ (b) $B = \{1,2,3,4,5\}$ (c) $C = \{4,2,3,5,1\}$
   (d) $D = \{b,a,d,c,e\}$ (e) $E = \{a,5,3,b,4\}$ (f) $F = \{b,3,a,5,4\}$

1.9 Intersection and Union of Sets

We denote the intersection of sets by ∩. Thus, we read $A \cap B$ as 'A intersection B' or 'the intersection of A and B' or 'A cap B'. The intersection of two sets A and B is the set, which consist of all the elements that are common to both set A and set B. i.e. $A \cap B = \{x : x \in A \text{ and } x \in B\}$.

If there are many sets, the intersection of these sets will be the set that consist of all the elements common to all these sets.

We denote the union of sets by ∪. Thus, $A \cup B$ is read 'A union B' or the union of 'A and B' or A cup B'. The union of two sets A and B is a set containing all the elements found in set A and all the elements found in set B, put together.

If there are, many sets the union of all these sets is a set containing all the elements of all these sets put together.

Example

$\mathscr{E}$ = {0,1,2,3,4,5,6,7,8,9,10,11,12}
A = {2, 4, 6, 8, 10, 12}
B = {1, 2, 3, 6, 9}
Find (a) A∩B (b) A∪B

Solution
(a) A∩B = {2,6} (b) A∪B = {1, 2, 3, 4, 6, 8, 9, 10, 12}

1.10 Disjoint Sets

Disjoint sets are sets which have no elements in common. If two sets A and B are disjoint, we can express this symbolically as $A \cap B = \emptyset$ or $n(A \cap B) = 0$.

1.11 Representation of Sets - Venn Diagrams

Venn diagrams are diagrams invented by an English Mathematician John Venn Euler (1834-1923) to represent relationships between sets. The usual way is to use a large rectangle to represent the universal set and circles to represent its subsets, (though we can use any other plane figure)

The following are some Venn diagrams representing different relationships between given sets.

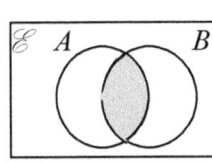
A∩B is shaded

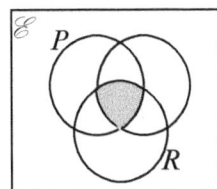
P∩Q∩R is shaded

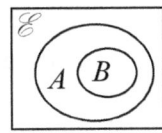
B ⊂ A

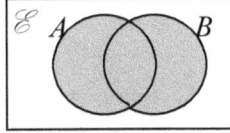
A∪B is shaded

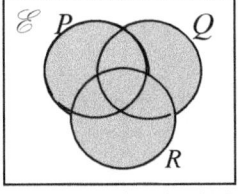
P∪Q∪R is shaded

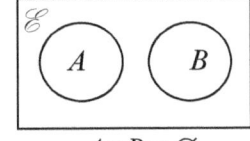
A∩B = ∅
Disjoint sets

Example

1. $\mathscr{E} = \{1,2,3,4,5,6,7,8,9,10\}$
   $A = \{1, 3, 4, 5, 7\}$
   $B = \{3, 4, 8, 9\}$

Draw a Venn diagram to illustrate the relationship between these sets. Using this Venn diagram, list the elements of $A \cap B$.

Solution

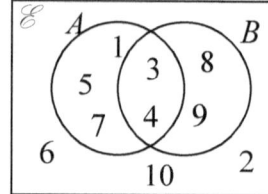

From the Venn diagram, $A \cap B = \{3, 4,\}$

     1.

2. $\mathscr{E} = \{1,2,3,4,5,6,7,8\}$
   $A = \{1,2,3,4\}$
   $B = \{4,6,8\}$

Draw a Venn diagram to show the relationship between these sets, listing the elements in each region. Using this Venn diagram, find $A \cup B$.

Solution

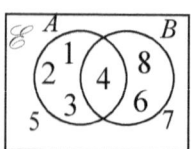

From the Venn diagram above, $A \cup B = \{1,2,3,4,6,8\}$.

3. $\mathscr{E}$ = Students of form one
   A = All form one boys
   B = All form one girls

State whether the sets A and B are disjoint.

Solution

A and B are disjoint.

Module 10, Topic 1: Sets

1.12 The Complement of a Set $C_\mathscr{E}^A$ or A'

The complement of a set A is denoted by A' or $C_\mathscr{E}^A$ and is the set containing all the elements of the universal set $\mathscr{E}$ that do not belong to the set A.

 Example

Let $\mathscr{E}$ = Whole numbers from 1 to 10 and $B = \{2,4,6,8,10\}$. Find B'.

Solution
$B' = \{1, 3, 5, 7, 9\}$

The following Venn diagrams represent the complement of sets. The shaded portion is that described under each Venn diagram.

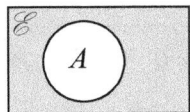

The shaded portion is A'

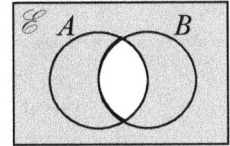

The shaded portion is
$(A \cap B)' = A' \cup B'$

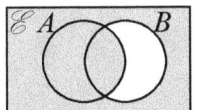

The shaded portion is
$A \cup B'$

The shaded portion is
$A \cap B'$

The shaded portion is
$(A \cup B)' = A' \cap B'$

The shaded portion is
$(A \cap B) \cup (A \cup B)'$

1.13 The Power or Derived Set of a Set

The power or derived set $P(A)$ of a set A is a set, which consists of all the subsets of A.

 Example

1. If $A = \{a, b\}$, find the power set of A.

   Solution
   $P(A) = \{\emptyset, \{a\}, \{b\}, \{a, b\}\} = \{\emptyset, \{a\}, \{b\}, A\}$

2. Find the derived set of $B = \{1, 3, 5\}$

   Solution
   $P(B) = \{\emptyset, \{1\}, \{3\}, \{5\}, \{1,3\}, \{1,5\}, \{3,5\}, B\}$

 Investigative Activity

Given that (i) $\emptyset = \{\}$ (ii) $A = \{a\}$ (iii) $B = \{a, b\}$ (iv) $C = \{a, b, c\}$.
1. Find the power set of each of the following sets.
2. State the cardinality of each of the power sets.
3. Write down a formula which you can use to find the cardinality of the power sets of any given set.

From the above investigation, we can see that:

$n(\emptyset) = 0$ and $n(P(\emptyset)) = 1 = 2^0$

$n(A) = 1$ and $n(P(A)) = 2 = 2^1$

$n(B) = 2$ and $n(P(B)) = 4 = 2^2$

$n(C) = 3$ and $n(P(C)) = 8 = 2^3$

Hence, $n(Y) = x \iff n(P(Y)) = 2^x$

Thus,

The cardinality of the power or derived set of a set with n elements is 2^n.

Careful observations of the above power or derived sets reveal that each power set contains the improper subsets $\emptyset$ and the set itself.

Therefore, if only the proper subsets are considered, the cardinality of the power set of a set with n elements will be $2^n - 2$.

Module 10, Topic 1: Sets

Exercise 1:4

1. Describe the shaded region in each of the following Venn diagrams using set notation.

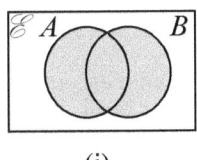 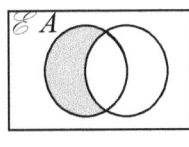 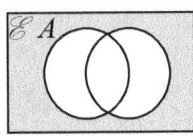
   (i)   (ii)   (iii)

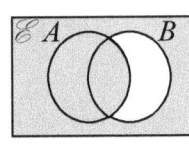   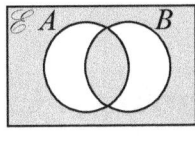
   (iv)   (v)

2. Given that A and B are subsets of $\mathscr{E}$ the universal set. Shade on different Venn diagrams
   (i) A (ii) A' (iii) $A \cap B$ (iv) $(A \cap B)'$
   (v) $A \cap B'$ (vi) $A \cup B$ (vii) $(A \cup B)'$ (viii) $A \cup B'$

3. Represent the following on a Venn diagram
   (i) A and B have no elements in common. (ii) All the elements in A are in B.

4. Use only set notation to write down the following statements:
   (i) A and B have no elements in common. (ii) All the elements in A are in B.

5. Represent the following on a Venn diagram
   $\mathscr{E} = \{0,1,2,3,4,5,6,7,8,9\}$
   $A = \{2, 4, 6, 8\}$
   Where, $\mathscr{E}$ is the universal set.

6. Let $\mathscr{E}$ = letters of the English alphabet
   $X = \{a, b, c, d, e, f, g, h, i, j\}$
   $Y = \{a, e, i\}$
   Show by use of a Venn diagram, the relationship between $\mathscr{E}$, X and Y

7. Given that,
   $\mathscr{E} = \{0,1,2,3,4,5,6,7,8,9,10\}$
   $P = \{2, 4, 6, 8, 10\}$
   $Q = \{1, 2, 3, 6, 9\}$
   Represent $\mathscr{E}$, P and Q on a Venn diagram.

8. If $\mathscr{E}$ = Natural numbers from 1 to 20
   A = factors of 18
   B = Multiples of 3
   C = Prime numbers

Draw a Venn diagram to represent the relationship between these sets. Hence, find
(a) $A \cap B \cap C$ (b) $A \cap B$ (c) $B \cap C$ (d) $A \cap C$

9. $\mathscr{E}$ = Whole numbers greater than zero but less than or equal to 124.
   A = Multiples of 10
   B = Multiples of 15
   C = multiples of 25
   (a) Draw a Venn diagram to represent the relationship between these sets.
   (b) Hence, find $n(A \cup B)$ and $n(A \cap B \cap C)$.

1.14 Algebraic Laws of Sets

Under the operations of union, intersection and complement, sets obey the following laws, which we refer to as the **algebraic laws of sets**.

1.	Idempotent Laws	$A \cup A = A;\ A \cap A = A$
2.	Associative Laws	$A \cup (B \cup C) = (A \cup B) \cup C$ $A \cap (B \cap C) = (A \cap B) \cap C$
3.	Commutative Laws	$A \cup B = B \cup A;\ A \cap B = B \cap A$
4.	Distributive Laws	$A \cup (B \cap C) = (A \cup B) \cap (A \cup C)$ $A \cap (B \cup C) = (A \cap B) \cup (A \cap C)$
5.	De-Morgan's Laws	$(A \cup B)' = A' \cap B'$ $(A \cap B)' = A' \cup B'$
6.	Complement Laws	$A \cup A' = \mathscr{E};\ A \cap A' = \varnothing$ $\mathscr{E}' = \varnothing;\ \varnothing' = \mathscr{E};\ (A')' = A$
7.	Identity Laws	$A \cup \varnothing = A;\ A \cap \varnothing = \varnothing$ $A \cup \mathscr{E} = \mathscr{E};\ A \cap \mathscr{E} = A$

1.15 Cardinality Logic

Example

1. In a certain class, 10 students do both arts and science, 24 do science and 22 do arts. Given that every student does at least one of these options, how many students are there in the class?

Solution

Let $\mathscr{E}$ = All the students
S = Science students
And A = Arts students

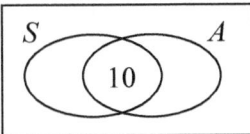 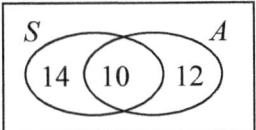

$n(\mathscr{E}) = 14 + 10 + 12 = 36$
Therefore, there are 36 students in the class.

2. During a conference attended by 30 participants, 18 of them ate rice, 13 ate beans and 5 of them ate both rice and beans. How many ate neither rice nor beans?

Solution

If $\mathscr{E}$ = All participants at a conference
R = Participants who ate rice
B = Participants who ate beans

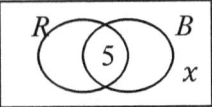 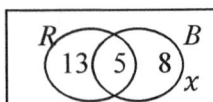

$x = n(R' \cap B')$
$n(\mathscr{E}) = n(R) + n(B) - n(R \cap B) + n(R' \cap B')$
$\Leftrightarrow 30 = 18 + 13 - 5 + n(R \cup B)'$
$\therefore n(R \cup B)' = 4$

Therefore, 4 participants ate neither beans nor ice.

3. A newsagent sells three papers: the Post, the Messenger and the Standard. Customers buy one of the three papers. Given that:
    80 customers buy the Post,
    70 customers buy the Messenger,
    60 customers buy the Standard,
    21 customers buy the Post and the Messenger,
    14 customers buy the Messenger and the Standard,
    16 customers buy the Standard and the Post,
    6 customers buy all three newspapers.
Draw a Venn diagram to illustrate this information. Hence, find the number of customers who buy only
    (a) The Post.   (b) The Messenger.   (c) The standard.
    (d) Find also the total number of customers for the three papers.

Solution

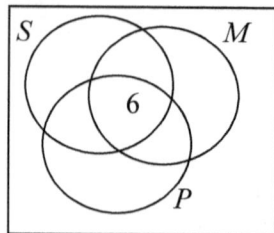
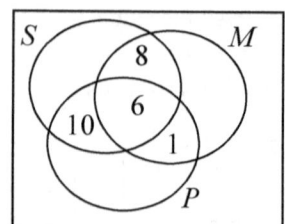

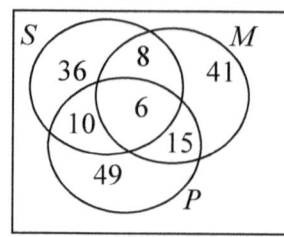

(a) Post customers only = $n((M \cup S)') = 49$.
(b) Messenger customers only = $n((P \cup S)') = 41$.
(c) Standard customers only = $n((M \cup P)') = 36$.
(d) Total number of customers = $n(M \cup P \cup S)$
   $= 36 + 41 + 49 + 8 + 10 + 15 + 6 = 144$

4. Out of 200 students in a school, 130 go for holidays in Douala and 120 go for holidays in Yaounde. Calculate the number of students who go to:
   (i) Both Douala and Yaounde, (ii) Douala only, (iii) Yaounde only,

   Solution
   $\mathscr{E}$ = All the 120 students
   D = those who go to Douala,
   Y = those who go to Yaounde
   Suppose x students go to both Douala and Yaounde.

   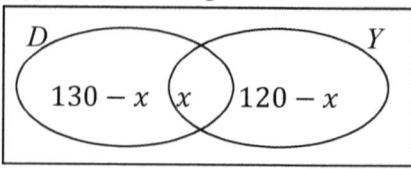

   $n(\mathscr{E}) = 200 = n(D \cap Y) + n(D') + n(Y')$
   $\Rightarrow 200 = x + (130 - x) + (120 - x)$
   $200 = 250 - x \Rightarrow x = 50$

Module10, Topic 1: Sets

 Exercise 1:5

1. The following information shows the number of people who visit a restaurant, which serves beans (*B*), rice (*R*), and achu (*A*).
   5 people did not eat any of the items served.
   Everyone who ate beans ate rice.
   3 people ate all the three items.
   10 people ate beans.
   4 people ate only rice
   20 people ate only achu
   10 people ate rice and not beans
   By using a Venn diagram or otherwise, find the number of people who
   (a) Ate achu (b) ate rice and beans but not achu
   (c) Visited the restaurant.

2. Out of a group of 20 persons, 7 like njang music, 12 like makossa music and 10 like bikutsi music; furthermore, 3 like both njang and bikutsi, 2 like both njang and makossa and 2 like all three kinds of music. Draw a Venn diagram and find how many of the 20 persons like makossa and bikutsi but not njang. Assume all the 20 persons like at least one of the three kinds of music.

3. A universal set $\mathscr{E}$ includes subsets *A*, *B* and *C*. There are 5 members of $\mathscr{E}$, who are not in any of these subsets. Every member of *B* is a member of *A* but *C* contains members who do not belong either to *A* or *B*. Draw a Venn diagram to incorporate these features.
   Given further that,
   (a) 20 members belong to *C* but not to *A* or *B*,
   (b) 3 members belong to *A*, *B* and *C*,
   (c) 13 members belong to *B*,
   (d) 27 members belong to *A* but not to *C*,
   (e) 10 members belong to both *A* and *C*.
   Insert appropriate numbers in the regions of your diagram. Hence, calculate,
   (i) $n(A)$, (ii) $n(A \cap C \cap B')$, (iii) $n(\mathscr{E})$.

4. Out of 93 students in a lower sixth class, 56 offer History and 59 offer geography. Only 8 students offer History but not geography. Find how many of the students offer
   (a) Geography but not History (b) Neither Geography nor History.

1.16 Transcribing from Set Language to Ordinary Language

The ability to transcribe from ordinary language to set language and vice versa is a very interesting exercise and useful tool to mathematicians of all levels. Initially this may appear to be very challenging but after practice this is very easy.

The following words are worth noting when transcribing from set language to ordinary language and vice versa.

1. The word **all** is used when one set is a subset of another set.
2. The word **some** is used when the intersection of two or more sets is not empty.
3. The word **none** is used when the intersection of two or more sets is empty.
4. Other words which are used include **and** which stands for intersection, **either…or** which stands for union, **neither nor** which stands for disjoint etc.

 Example

Let M = ministers, R = Rav4 owners, L = large salary earners, S = smugglers. Use words such as some, all, none, both, and the Venn diagram below to construct sentences to interpret the following.
(a) $M \subset L$
(b) $(S \cap L) \cap M \neq \emptyset$
(c) $(M \cap R) \subset L$
(d) $(M' \cap R) \cap L \neq \emptyset$

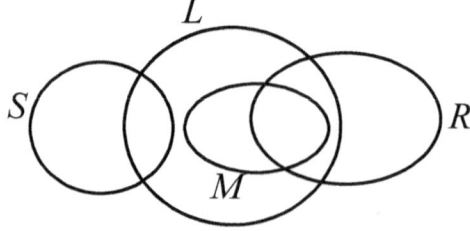

Solution
(a) All ministers are large salary earners.
(b) None of the large salary earners who are smugglers are ministers.
(c) All ministers who ware Rav4 owners are large salary earners.
(d) Some large salary earners who are Rav4 owners are not ministers.

 Exercise 1:6

1. Consider the following statements
   A = The illegal sale of drugs is rising.
   B = The customs department is concerned with rising drug traffic

Describe in words the statements represented by the following set operations.
(a) $A \cap B$ (b) $A \cup B$ (c) B' (d) $A \cap B'$ (e) $A' \cap B'$

2. Given that $\mathscr{E}$ = all men, H = honest men, B = businessmen, S = successful men.
   (a) Using the symbols for union, intersection, or complement express in set notation the set of dishonest, unsuccessful businessmen.
   (b) On the Venn diagram below, shade the region which represents the set of honest, unsuccessful businessmen.

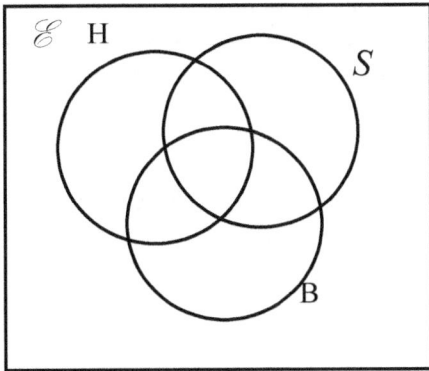

3. M = mammals, V = vertebrates, I = insects, H = horned animals. Write in set notation using the symbols for union, intersection or subsets.
   (a) All mammals are vertebrates. (b) No insect is a vertebrate.
   (c) Some mammals are horned animals.

4. Each of the students leaving a certain school registered for at least one of the following competitive examinations Police constable, teacher's grade 1, and Nursing aid. Taking

   $\mathscr{E}$ = students leaving the school.
   P = students who registered for Police constable.
   T = students who registered for teacher's grade 1.
   N = students who registered for nursing aid.

   Express the following statement of facts in set language.
   (a) All the students registered at least one of the competitive examinations.
   (b) All those who registered for both Police constable and nursing aid, also registered for teachers grade one.
   (c) In no case was Police constable the only competitive examination to which registration was made.

Multiple Choice Exercise 1

1. Let $P = \{2, 4, 6, 8, 10\}$, $Q = \{1, 3, 5, 7\}$, $R = \left\{\frac{1}{2}, \frac{1}{4}, \frac{1}{6}, \frac{1}{8}\right\}$ and $S = \{10, 6, 2, 8, 4\}$. It is true to say that:
   [A] P and Q are equivalent sets.   [B] Q and R are equivalent sets
   [C] P and S are equivalent sets   [D] Q and R are equal sets

2. Let $P = \{2, 4, 6, 8, 10\}$, $Q = \{1, 3, 5, 7\}$, $R = \left\{\frac{1}{2}, \frac{1}{4}, \frac{1}{6}, \frac{1}{8}\right\}$ and $S = \{10, 6, 2, 8, 4\}$. It is not true to say that:
   [A] P and Q are equivalent sets   [B] Q and R are equivalent sets
   [C] P and S are equal sets   [D] P and P∩S are equal sets.

3. Given that $H = \emptyset$. It is true to say that:
   [A] $0 \in H$ [B] $n(H) = \emptyset$ [C] $H = \{0\}$ [D] $n(H) = 0$

4. Given the universal set $\mathscr{E} = \{x : 0 < x < 10, x \in Z\}$. The complement of the set $P = \{x : x \in \mathscr{E}, x \text{ is not divisible by } 4\}$ is:
   [A] {4}   B] {4,8}   [C] {1,2,3}   [D] {1,2,3,5,6,7,9}

5. If $A = \{a, b, c\}$, $B = \{a, b, c, d, e\}$ and $C = \{a, b, c, d, e, f\}$, $(A \cup B) \cap (A \cup C)$ is equal to:
   [A] {a, b, c, d}   [B] {a, b, c, d, e}   [C] {a, b, c, d, e, f}   [D] {a, b, c}

6. Let J be the set of positive integers. If $H = \{x : x^2 < 3, x \neq 0\}$, then:
   [A] $H = \{1\}$ [B] H is an infinite set   [C] $H = \{0, 1\}$ [D] $H = \{\}$

7. Given that $P = \left\{2, 1, 3, 9, \frac{1}{2}\right\}$, $Q = \left\{1, 2\frac{1}{2}, 3, 7\right\}$, $R = \left\{5, 4, 2\frac{1}{2}\right\}$ and $P \cup Q \cup R$ is equal to:
   [A] $\left\{5, 4, 2\frac{1}{2}\right\}$ [B] {1,2,3,4,5,6,7}   [C] {1,9}   [D] $\left\{\frac{1}{2}, 1, 2, 2\frac{1}{2}, 3\ 4\ 5\ 7\ 9\right\}$

8. Given that $P = \left\{2, 1, 3, 9, \frac{1}{2}\right\}$, $Q = \left\{1, 2\frac{1}{2}, 3, 7\right\}$, $R = \left\{5, 4, 2\frac{1}{2}\right\}$ and $P \cap Q \cap R$ is equal to:
   [A] {5,7,9}   [B] ∅   [C] {1,3,7}   [D] {4}

9. Let $\mathscr{E} = \{1,2,3,4\}$, $P = \{2,3\}$ and $Q = \{2,3,4\}$. P∩Q is equal to:
   [A] {1,2,3}   [B] {1,3,4}   [C] {2,3}   [D] {1,3}

10. $S = \{1,2,3,4,5,6\}$, $T = \{2,4,5,7\}$ and $R = \{1,4,5\}$. Then $(S \cap T) \cup R$ is:
    [A] {1,4,5}   [B] {2,4,5}   [C] {2,3,4,5}   [D] {1,2,4,5}

11. If $R = \{2,4,6,7\}$ and $S = \{1,2,4,8\}$ then R∪S equals:
    [A] {1,2,4,6,7,8}   [B] {1,2,4,7,8}   [C] {1,4,7,8}   [D] {2,6,7}

12. If $P = \{3,5,6\}$ and $Q = \{4,5,6\}$ then P∩Q equals:
    [A] {3,6}   [B] {4,5}   [C] {4,6}   [D]{5,6}

13. If $S \subset R$, then:
    [A] $S \cap R = R$ [B] $S \cap R = S$ [C] $S \cup R = S'$ [D] $S \cup R = S$

14. If $P = \{3,7,11,13\}$, $Q = \{2,4,8,16\}$. It follows that:

[A] $(P \cap Q)' = \{2,3,4,13\}$ [B] $n(P \cup Q) = 4$ [C] $P \cup Q = \emptyset$ [D] $P \cap Q = \emptyset$

15. Given that $P = \{b, d, e, f\}$ and $Q = \{a, c, f, g\}$ are subsets of the universal set $U = \{a, b, c, d, e, f, g\}$. $P' \cap Q$ is equal to:
    [A] $\{a, c\}$ [B] $\{a, c, d, g\}$ [C] $\{c, d, g\}$ [D] $\{a, c, g\}$

16. In the Venn diagram above, the shaded region is described by
    [A] $Q \cap R$ [B] $Q \subset R$ [C] $P \cap Q \cup R$ [D] $P \cup Q \cap R$

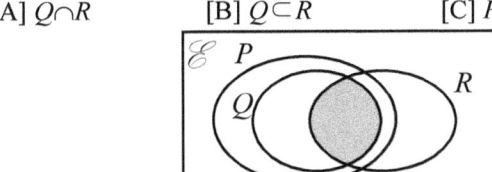

17. Given the universal set $\mathscr{E} = \{1,2,3,4,5\}$, $P = \{1,2\}$, $Q = \{2,3,4\}$, then $(P \cap Q)'$ is
    [A] $\{2\}$ [B] $\{5\}$ [C] $\{1,3,4,5\}$ [D] $\{1,2,3,5\}$

19. The figure above, given that $\dfrac{4}{9}$ of the total number of students offering Physics (P) or Biology (B) offer Physics only. The number of students offering both subjects is:
    [A] 27   [B] 20   [C] 7   [D] 18

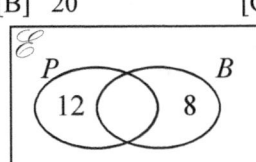

19. In the following figure, $n(P \cap Q)$ is:
    [A] 1   [B] 2   [C] 4   [D] 6

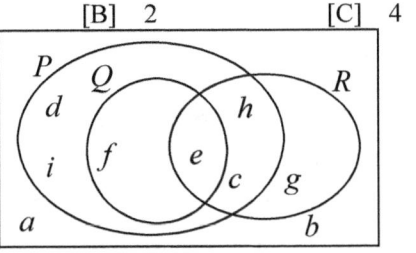

20. In the figure above, $Q' \cap R$ is equal to:
    [A] $\{e\}$ [B] $\{c, h\}$ [C] $\{c, g, h\}$ [D] $\{c, e, g, h\}$

21. In the figure above, the shaded portion is:
    [A] $P' \cap Q$ [B] $(R \cap Q) \cup (P' \cap R)$ [C] $P' \cap Q \cap R$ [D] $(P \cup Q)' \cap R$

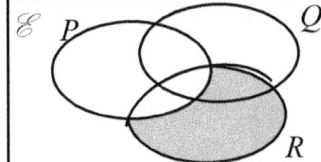

22. In a class of 80 students, every student study Economics or Geography. If 65 students study Economics and 50 study Geography, the number of students who study both subjects is:
    [A] 15 [B] 30 [C] 35 [D] 45
23. A and B are two sets. The number of elements in $A \cup B$ is 49. The number in A is 22 and the number in B is 34. The number of elements in $A \cap B$ is:
    [A] 7 [B] 27 [C] 15 [D] 12
24. If $n(P) = 21$, $n(R) = 33$ and $n(P \cup R) = 46$. $n(P \cap R)$ is equal to:
    [A] 8 [B] 34 [C] 58 [D] 100
25. The Venn diagram below shows the number of students who studied physics P, chemistry C, and mathematics M in a certain school. How many students studied at least two of the three subjects?
    [A] 165 [B] 160 [C] 155 [D] 135

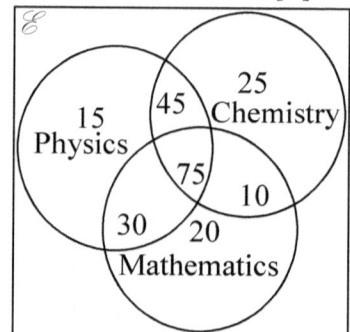

26. Let $\mathscr{E}$ = All men, H = honest persons, B = Businesspersons, S = Successful persons. We can represent the set of dishonest, unsuccessful businesspersons by:
    [A] $H \cap S' \cap B$ [B] $H \cap S \cap B$ [C] $H' \cup S' \cup B$ [D] $H' \cap S' \cap B$
27. Let M = mammals, V = vertebrates, I = insects, H = horned animals. We can denote the statement, 'All mammals are vertebrates' by:
    [A] $M \cap V \neq \emptyset$ [B] $M \subset V$ [C] $M \cup V = \emptyset$ [D] $M \cup H \neq \emptyset$
28. Let M = mammals, V = vertebrates, I = insects, H = horned animals. $V \cap I = \emptyset$ in ordinary English means:
    [A] Vertebrates and insects are empty sets. [B] Some vertebrates are insects
    [C] Vertebrates cannot join with insects. [D] No insect is a vertebrate.
29. Let M = mammals, V = vertebrates, I = insects, H = horned animals. Using set notation we can write the statement 'Some mammals are horned animals' as:
    [A] $H \cap M \neq \emptyset$ [B] $M \cap H = \emptyset$ [C] $M \cup H \neq \emptyset$ [D] $M \cup H = \emptyset$
30. The statement $A \subset B$ is the same as:
    [A] $A \cap B = B$ [B] $A \cap B = A$ [C] $A \cup B = B$ [D] $A \cup B = A$

Topic 2

INDICES AND LOGARITHMS

Objectives

At the end of this topic, the learner should be able to:

1. State and apply laws of indices.
2. Solve simple equations involving indices.
3. State and apply laws of logarithms to simple logarithmic expressions and equations.
4. Perform operations in bases other than ten.
5. Find the values of numbers given in index form or in logarithmic form.
6. Solve simple logarithmic equations.

INTRODUCTION TO INDICES

2.1 Meaning of b^p

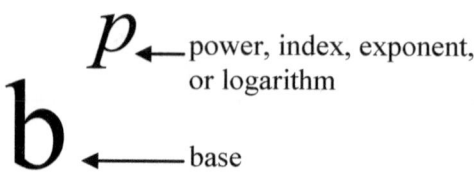

We studied the meaning of the index notation in Topic 2. Just to recall, another name for **power** is **index, exponent** or **logarithm**.

2.2 Law of Indices

If a is a non-zero real number and m and n are integers then:
1. **Exponent 1 Law:** $a^1 = a$.
2. **Exponent 0 Law:** $a^0 = 1$.
3. **Multiplication Law:** $a^m \times a^n = a^{m+n}$
4. **Division Law:** $a^m \div a^n = a^{m-n}$
5. **Negative Index Law:** $a^{-n} = \dfrac{1}{a^n}$

 Example

Evaluate the following, allowing your answer in index form.

(a) $5^2 \times 5^4 \times 5^1 \times 5^3$ (b) $b^4 \times b^2 \times b^3 \times b^0$ (c) $81(x)^9 \div (3x)^3$

(d) $\dfrac{(2p)^3 \times (3p)^2}{36p^4}$ (e) $16x^5 \times 3x^3 \div 96x^8$ (f) $4y^6 \times 9y^{-10} \div 72y^{-3}$

Solution
(a) $5^2 \times 5^4 \times 5^1 \times 5^3 = 5^{2+4+1+3} = 5^{10}$

(b) $b^4 \times b^2 \times b^3 \times b^0 = b^{4+2+3+0} = b^9$

(c) $81(x)^9 \div (3x)^3 = (3^4)(x)^9 \div (3^3)x^3 = (3^{4-3})(x^{9-3}) = 3x^6$

(d) $\dfrac{(2p)^3 \times (3p)^2}{36p^4} = \dfrac{2^3 \times p^3 \times 3^2 p^2}{4 \times 9 p^4} = \dfrac{2^3 \times p^3 \times 3^2 p^2}{2^2 \times 3^2 p^4} = 2^{3-2} \times 3^{2-2} \times p^{3+2-4} = 2p$

(e) $16x^5 \times 3x^3 \div 96x^8 = (16 \times 3)x^{5+3} \div 96x^8 = \dfrac{48x^{8-8}}{96} = \dfrac{1}{2}x^0 = \dfrac{1}{2}$

(f) $4y^6 \times 9y^{-10} \div 72y^{-3} = \dfrac{4 \times 9 \times y^{6-10}}{72y^{-3}} = \dfrac{y^{6-10+3}}{2} = \dfrac{1}{2y}$

Exercise 2:1

Evaluate the following.
1. 6^3
2. 8^3
3. $2^3 \times 3^2$
4. $3^3 \times 5^2$
5. $y^3 \times y^2 \times y^5$
6. $x^2 \times 4x^5 \times 2x^3$
7. $3x \times 4x^3 \times 5x^5$
8. $\dfrac{a^2 b^4 c^7}{c^7 a^2 b^4}$
9. $\dfrac{10a^6}{5a^3}$
10. $\dfrac{2x^2 y}{3xy^2}$
11. $\dfrac{12a^2}{18a^2 b}$
12. $a^4 \div a^6$
13. $25x^5 \div 5x^4$
14. $20a^3 b^5 \div 5a^2 b^3$
15. $(a^2 b^5) \div (a^3 b^4)$
16. $\dfrac{-5x^2}{15x}$
17. $16a^5 \div 8a^3$
18. $\dfrac{5x^2}{x}$

The product index law

 Investigative Activity

1. Evaluate (a) 2^3 (b) $(2^3)^2$ (c) 2^6.
2. Evaluate (a) 3^2 (b) $(3^2)^3$ (c) 3^6.
3. Deduce an expression for the value of $(a^m)^n$ in index form.

From the above investigation, we can conclude that $(a^m)^n = a^{mn}$

Example

Simplify $\left(\dfrac{2a^2b^4}{c^3}\right)^2$.

Solution

$$\left(\dfrac{2a^2b^4}{c^3}\right)^2 = \dfrac{4a^4b^8}{c^6}$$

The fractional index laws

(a) $\quad b^{\frac{1}{n}} = \sqrt[n]{b}$ (b) $\quad b^{\frac{m}{n}} = \left(\sqrt[n]{b}\right)^m = \sqrt[n]{b^m}$

Example

Simplify $\left(\dfrac{625}{144}\right)^{-\frac{3}{2}}$

Solution

$$\left(\dfrac{625}{144}\right)^{-\frac{3}{2}} = \left(\dfrac{144}{625}\right)^{\frac{3}{2}} = \left(\dfrac{12}{25}\right)^3 = \dfrac{1728}{15625}$$

Product of Numbers with the same Power

Investigative Activity

1. Evaluate (a) $3^2 \times 4^2$ (b) $(3 \times 4)^2 = 12^2 = 144$
2. Evaluate (a) $2^3 \times 4^3$ (b) $(2 \times 4)^3$
3. What conclusion do you draw?

From the above investigation, we can conclude that

$$\therefore a^m b^m = (ab)^m \quad \text{............(9)}$$

To multiply numbers with a common power, multiply the numbers and retain their common power.

2.3 Exponential or Index Equations

We call equations involving indices **exponential or index equations**. To solve **exponential or index equations** we apply the laws of indices at various stages.

 Example

Find the value of x for which $64^x = 16^{2x+1}$.

Solution

$$64^x = 16^{2x+1}$$

Since $64 = 2^6$ and $16 = 2^4$, we can express the equation in terms of the common base 2. Thus,

$$2^{6x} = 2^{4(2x+1)}$$

Equating exponents

$$6x = 4(2x + 1) \Rightarrow 6x = 8x + 4$$
$$-2x = 4 \Rightarrow x = -2$$

 Exercise 2:2

1. Simplify the following.

   (a) $64^{\frac{1}{3}}$ (b) $\left(\dfrac{16}{25}\right)^{-\frac{1}{2}}$ (c) $\dfrac{5^3 \times 5^0}{25}$ (d) $\left(216^{\frac{1}{3}}\right)^{-2}$ (f) $6^{-3} \times 2^5 \times 3^3$

2. Solve the following equations.

   (a) $4^x = 2^{\frac{1}{2}} \times 8$ (b) $9 \times 3^{3+x} = 27^{-x}$ (c) $3 \times 9^{1+x} = 27^{-x}$

   (d) $x^4 = (0.25)^2$ (e) $9^{2x+1} = \dfrac{81^{x-1}}{3^x}$ (f) $\dfrac{9^{2x-1}}{3^{x+3}} = 1$

   (g) $9^x = 729$ (h) $9^{2x-1} \times 3^{x+1} = 27^{x+3}$ (i) $27^{2x+1} = 3^{x-1}$

   (j) $2^{x-1} = 32$ (k) $3^{2y} = \dfrac{1}{27}$ (l) $8^{x-1} = 16$

31

THEORY OF LOGARITHMS

2.4 Definition of Logarithms

The logarithm of a positive number n to a base b is the power p to which we must raise b to give the number n.

Using algebraic notation;

$$\log_b n = p \Leftrightarrow n = b^p, \; n > 0, \; n \in \Box$$

We can use the above definition to transform logarithmic equations to exponential equations and vice versa.

Note that the logarithm of negative numbers and zero do not exist.

Example

1. Write down the following in logarithmic form
   (a) $3^4 = 81$ (b) $x^y = z$ (c) $(ab)^{pq} = xz$

   Solution
   (a) $\log_3 81 = 4$ (b) $\log_x z = y$ (c) $\log_{(ab)} xz = pq$

2. Write down the following in exponential form
   (a) $a = \log_b c$ (b) $\log_{10} 100 = 2$ (c) $\log_{x^2} 3y = n$
   Solution
   (a) $b^a = c$ (b) $10^2 = 100$ (c) $(x^2)^n = 3y$ or $x^{2n} = 3y$

3. Use the definition of logarithms to simplify the following.
   (i) $\log_{10} 100$ (ii) $\log_{10} 10000$ (iii) $\log_b 1$

Solution
(i) Let $\log_{10} 100 = x \Leftrightarrow 10^x = 100$
$$10^x = 10^2 \Rightarrow x = 2$$
(ii) Let $\log_{10} 10000 = x \Leftrightarrow 10^x = 10000$
$$10^x = 10^4 \Rightarrow x = 4$$
(iii) Let $\log_b 1 = n \Leftrightarrow b^n = 1$.
Since for any real number b, $b^0 = 1$ then $n = 0 \Rightarrow \log_b 1 = 0$

The Logarithm of 1

$$\log_b 1 = 0$$

The logarithm of 1 to any base is zero.

2.5 Logarithmic Equations

Logarithmic equations are equations, which involve logarithms. The definition of logarithms is very useful in solving many logarithmic equations.

Example

Use the definition of logarithms to solve the equation $\log_8 x = 3$.

Solution
$\log_8 x = 3 \Rightarrow x = 8^3 = 512$

Exercise 2:3

(i) Evaluate the following.

1. $\log_x x^n$ 2. $\log_8 64$ 3. $\log_5 \left(\dfrac{1}{5}\right)$ 4. $\log_5 125$

5. $\log_{125} 5$ 6. $\log_3 81$ 7. $\log_3 243$ 9. $\log_4 8$

(ii) Express the following as exponential equations.
     9. $\log_n y = 3$ 10. $\log_{10} x = 3$

(iii) Find the value of the unknown in each of the following.
     11. $\log_4 n = 0$ 12. $\log_{10} n = 1$ 13. $\log_a y = 0$
     14. $\log_4 x = 2$ 15. $\log_{10} x = -1$ 16. $\log_3 81 = x + 1$

2.6 Laws of Logarithms

1. *The Addition Law of logarithms*

$$\log_b xy = \log_b x + \log_b y$$

2. *The Subtraction Law of logarithms*

$$\log_b \left(\frac{x}{y}\right) = \log_b x - \log_b y$$

3. *The Exponential Law of Logarithms*

$$\log_b x^n = n \log_b x$$

 Example

Without using tables or calculators, simplify the following.

1. $\log_{10} 8 + \log_{10} 125$ 2. $\log_3 21 - \log_3 7$ 3. $\log_5 0.25 + \log_5 100$ 4. $\dfrac{\log_{10} 81}{\log_{10} 27}$

5. $\log_{10} 75 + 2\log_{10} 2 - \log_{10} 3$ 6. $\log_{10} 64 + 2\log_{10} 5 - 2\log_{10} 40$

Solutions

1. $\log_{10} 8 + \log_{10} 125 = \log_{10}(8)(125) = \log_{10} 1000 = 3$

2. $\log_3 21 - \log_3 7 = \log_3 \dfrac{21}{7} = \log_3 3 = 1$

3. $\log_5 0.25 + \log_5 100 = \log_5 (0.25)(100) = \log_5 25$

   $\Rightarrow \log_5 0.25 + \log_5 100 = \log_5 5^2 = 2\log_5 5 = 2$

4. $\dfrac{\log_{10} 81}{\log_{10} 27} = \dfrac{\log_{10} 3^4}{\log_{10} 3^3} = \dfrac{4 \log_{10} 3}{3 \log_{10} 3} = \dfrac{4}{3}$

5. $\log_{10} 75 + 2\log_{10} 2 - \log_{10} 3 = \log_{10}\left(\dfrac{75}{3}\right) + \log_{10} 2^2 = \log_{10}(25)(4) = \log_{10} 100 = 2$

6. $\log_{10} 64 + 2\log_{10} 5 - 2\log_{10} 40 = \log_{10} 64 + \log_{10} 25 - \log_{10} 1600$

   $= \log_{10} \dfrac{64(25)}{1600} = \log_{10} 1 = 0$

Exercise 2:4

(1) Simplify the following without using tables or calculators.

(a) $\dfrac{\log_{10} 16 - \log_{10} 2}{\log_{10} 2 - \log_{10} 1}$ (b) $\dfrac{\log_{10} 81 + \log_{10} 27}{\log_{10} 9 - \log_{10} 3}$ (c) $\log_{10} 125 + 8$

(d) $\log_{10} 30 - \log_{10} 3$ (e) $2\log_{10} 8 + 2\log_{10} 5 - 4\log_{10} 2$

(f) $\dfrac{\log_{10} a + \log_{10} a^2}{\log_{10} a}$ (g) $\log_{10} 15 + 3\log_{10} 2 - 4\log_{10} 1.2$

(h) $3\log_{10} 2 + \log_{10} 200 - \log_{10} 16$

Multiple Choice Exercise 2

1. The value of $(2^3)^2$ is:
   [A] 16   [B] 32   [C] 36   [D] 64
2. The value of $2^2 + 3^3$ is:
   [A] 13   [B] 25   [C] 31   [D] 36
3. The value of $64^{\frac{1}{3}}$ is:
   [A] 16   [B] 8   [C] 4   [D] 2
4. 5^4 has the value of is:
   [A] 9   [B] 20   [C] 125   [D] 625
5. The value of $2^0 - 2^{-2}$ is:
   [A] 1   [B] $-\frac{1}{4}$ [C] $\frac{3}{4}$ [D] $\frac{1}{4}$
6. $\dfrac{2^3 \times 2^4}{2^2}$ is equal to:
   [A] 2^4 [B] 2^3 [C] 2^2 [D] 2^5
7. The value of $\left(\dfrac{196}{225}\right)^{-\frac{1}{2}}$ is:
   [A] $\frac{17}{14}$ [B] $\frac{15}{14}$ [C] $\frac{14}{15}$ [D] $\frac{14}{17}$
8. On simplification $16^{\frac{1}{2}}(4^{-1} + 5^0)$ gives:
   [A] 5   [B] $5\frac{1}{2}$ [C] $4\frac{1}{2}$ [D] $4\frac{1}{4}$
9. $(0.7)^3$ equals:
   [A] 2.1   [B] 0.49   [C] 3.43   [D] 0.343
10. When simplified $(27^2)^{\frac{1}{3}}$ equals:
    [A] 81   [B] 6   [C] 9   [D] 8
11. After evaluating $36^{\frac{1}{2}} \times 64^{-\frac{1}{2}} \times 5^0$ equals:
    [A] $\frac{3}{4}$ [B] $\frac{1}{24}$ [C] $\frac{2}{3}$ [D] $1\frac{1}{2}$
12. When $\dfrac{9^{-\frac{1}{2}}}{27^{\frac{2}{3}}}$ is simplified the result is:
    [A] $\frac{1}{2}$ [B] $\frac{1}{9}$ [C] $\frac{1}{18}$ [D] $\frac{1}{27}$

13. Simplifying $125^{-\frac{1}{3}} \times 49^{-\frac{1}{2}} \times 10^0$ gives:

    [A] 350   [B] 35   [C] $\frac{1}{35}$ [D] $\frac{1}{350}$

14. $\left(\frac{1}{4}\right)^{-1\frac{1}{2}}$ when simplified is:

    [A] $\frac{1}{8}$ [B] $\frac{1}{4}$ [C] 8   [D] 4

15. On evaluation $\left(\frac{16}{81}\right)^{\frac{1}{4}}$ becomes:

    [A] $\frac{8}{27}$ [B] $\frac{1}{3}$ [C] $\frac{4}{9}$ [D] $\frac{2}{3}$

16. $16^{\frac{5}{4}} \times 2^{-3} \times 3^0$ is equal to:

    [A] 20   [B] 2   [C] 4   [D] 10

17. The result of evaluating $0.027^{-\frac{1}{2}}$ is:

    [A] $3\frac{1}{3}$ [B] 3   [C] $\frac{3}{10}$ [D] $\frac{1}{3}$

18. $\dfrac{8^{\frac{2}{3}} \times 27^{-\frac{1}{3}}}{64^{\frac{1}{3}}}$ simplifies to:

    [A] $\frac{1}{3}$ [B] $\frac{1}{9}$ [C] $\frac{16}{3}$ [D] $\frac{27}{8}$

19. After evaluating $5\frac{2}{3} \times \left(\frac{2}{3}\right)^2 \div \left(1\frac{1}{2}\right)^{-1}$ the result is:

    [A] $\frac{12}{5}$ [B] $\frac{8}{5}$ [C] $3\frac{3}{5}$ [D] $4\frac{1}{8}$

20. The result of simplifying $\dfrac{2^{\frac{1}{2}} \times 8^{\frac{1}{2}}}{4}$ is:

    [A] 1   [B] 2   [C] 4   [D] 16

21. $\left(\frac{1}{4}\right)^{-1\frac{1}{2}}$ is equal to:

    [A] 8   [B] 4   [C] $\frac{1}{4}$ [D] $\frac{1}{16}$

22. $(3a^2)^3$ is equal to:

    [A] $3a^6$ [B] $9a^6$ [C] $27a^2$ [D] $27a^6$

23. $a^2 \times b \times a^4 \times b^2$ simplifies to:

    [A] $a^6 b^2$ [B] $a^8 b^2$ [C] $a^3 b^3$ [D] $a^6 b^3$

Module 10, Topic 2: Indices and Logarithms

24. When simplified $\dfrac{x^3 y^4 z^7}{x^2 y^6 z^7}$ is equal to:

    [A] $\dfrac{x}{y^2}$ [B] $\dfrac{y^2}{x}$ [C] $\dfrac{x^2}{y}$ [D] $\dfrac{1}{y}$

25. $(m^2 n^5) \div (m^3 n^4)$ equals:

    [A] mn^{-1} [B] $m^{-1}n$ [C] $m^5 n^9$ [D] mn

26. When $10a^6$ is divided by $5a^3$ the result is:

    [A] $3a^3$ [B] $2a$ [C] a^3 [D] $2a^3$

27. If $2^x \times 3^2 = 144$, the value of x is:
    [A] 7   [B] 5   [C] 4   [D] 8

28. If $x^2 \times 3^2 \times 1^2 = 144$, the value of x is:
    [A] -4 [B] 2   [C] -2 [D] 16

29. When $3^x = 81$, the value of x is:
    [A] 2   [B] 3   [C] 4   [D] 27

30. The value of x for which $3^x = 243$ is:
    [A] 6   [B] 5   [C] 4   [D] 3

31. If $3^x + 6 = 87$, the value of x is:
    [A] 1   [B] 2   [C] 3   [D] 4

32. If $3^{2x} = 27$ the value of x is:
    [A] 1   [B] 1.5   [C] 4.5   [D] 18

33. Given that $27^{(1+x)} = 9$. The value of x is:

    [A] -3 [B] $-\dfrac{1}{3}$ [C] $\dfrac{1}{3}$ [D] 2

34. $\left(\dfrac{1}{4}\right)^{2-y} = 1$, the value of y is:

    [A] -2 [B] $-\dfrac{1}{2}$ [C] $\dfrac{1}{2}$ [D] 2

35. The solution of the equation $2\sqrt{x} = 4$ is:
    [A] -2 [B] 2   [C] 4   [D] 6

36. The value of x for which $2^{-6x} = 8^{(1-x)}$ is true is:

    [A] $-\dfrac{7}{3}$ [B] $\dfrac{1}{3}$ [C] -1 [D] $\dfrac{7}{9}$

37. When $56x^{-4} \div 14x^{-8}$ is simplified the result is:
    [A] $2x^{-12}$ [B] $4x^{-4}$ [C] $4x^{+4}$ [D] $4x^{-3}$

38. The word that is not another name for logarithm is:
    [A] Power   [B] exponent   [C] Base   [D] index

39. The value of $\log_{10} 6 + \log_{10} 45 - \log_{10} 27$ is:
    [A] 0 [B] 1 [C] 1.1738 [D] 10

40. Simplifying $\dfrac{\log 27^{\frac{1}{3}}}{\log 81}$ gives:
    [A] 14 [B] $\dfrac{3}{8}$ [C] $\dfrac{1}{2}$ [D] $\dfrac{3}{4}$

41. The value of $\log_{10} 25 + \log_{10} 32 - \log_{10} 8$ is:
    [A] 0.2 [B] 2 [C] 100 [D] 409

42. The value of $2\log_3 6 + \log_3 16$ is:
    [A] $4 - \log_3 2$ [B] $3 + \log_3 2$ [C] $2 + 6\log_3 2$ [D] $3 - \log_3 2$

43. On evaluation, $\log_{10} 4 + \log_{10} 25$ becomes:
    [A] 1 [B] 2 [C] 3 [D] 4

44. The value of $\log_{10} 5 + \log_{10} 20$ is:
    [A] 2 [B] 3 [C] 4 [D] 5

45. If $3\log_{10} a = \log_{10} 64$, the value of a is:
    [A] 4 [B] 6 [C] 8 [D] 16

46. $7^{x-1} = \log_5 5$, then x is equal to:
    [A] 1 [B] 7 [C] -1 [D] -7

47. Given that $\dfrac{1}{3}\log_{10} p = 1$. The value of p is:
    [A] 3 [B] 10 [C] 100 [D] 1000

48. If $\log_a x = p$, then in terms of a and p, x is equal to:
    [A] a^p [B] $\dfrac{a}{p}$ [C] p^a [D] ap

49. The value of p for which $\dfrac{1}{2}\log_{10} p = 1$ is true is:
    [A] 10^{-1} [B] 10^3 [C] 10^2 [D] 10^1

50. Given that $\log_4 x = -3$. The value of x is:
    [A] $\dfrac{1}{81}$ [B] $\dfrac{1}{64}$ [C] 64 [D] 81

Topic 3

MATRICES

Objectives

At the end of this topic, the learner should be able to:

1. Represent information in matrix form.
2. State the order of a matrix.
3. Identify some types of matrices.
4. Determine whether matrices are equivalent or equal.
5. Add and subtract matrices.
6. Multiply a matrix by a scalar.
7. Multiply a matrix by another matrix.
8. Find the transpose of a matrix.
9. Find the determinant of a 2 by 2 matrix.
10. Determine whether a matrix is singular.

3.1 Concept and Representation of Matrices

A trader records his stock for three days as follows:

   Monday   : 5 pens, 6 pencils, 2 rulers.

   Tuesday : 3 pens, 4 pencils, 1 ruler.

   Wednesday: 1 pens, 0 pencils, 3 rulers.

We can display this information in the following ways.

	Pens	Pencils	Rulers
Mon	5	6	2
Tues	3	4	1
Wed	1	0	3

Or simply

$$\begin{pmatrix} 5 & 6 & 2 \\ 3 & 4 & 1 \\ 1 & 0 & 3 \end{pmatrix}$$

The second display of this information is an example of a matrix.

3.2 Definition and Size of a Matrix

A matrix is a rectangular arrangement of numbers in rows r and columns c. The numbers are called the **elements** or **entries** of the matrix. Matrices are denoted by capital letters A, B, C etc while the elements are denoted by lower case letters $a, b, c,$...if at all they are letters. In printed text, bold type **A, B, C** etc are used.

The **size** or **order** of a matrix with r rows and c columns is specified as $r \times c$, read 'r by c'. The following matrix has 3 rows and 4 columns, so it is a 3×4 matrix.

$$\begin{pmatrix} 4 & 2 & 8 & 5 \\ 0 & 7 & 9 & 4 \\ -1 & 6 & 1 & 3 \end{pmatrix}$$

with Column 2 indicated (second column) and Row 2 indicated (second row).

Module 10, Topic 3: Matrices

Example

State the sizes of the following matrices.

(i) $\begin{pmatrix} 1 & 2 \\ 3 & 4 \end{pmatrix}$ (ii) $\begin{pmatrix} 2 & 3 \\ 4 & 6 \\ 7 & 5 \end{pmatrix}$ (iii) $\begin{pmatrix} 2 & 4 & 7 \\ 3 & 6 & 5 \end{pmatrix}$

Solution
(i) 2×2 (ii) 3×2 (iii) 2×3

The × in the notation for order of a matrix should not be confused with the multiplication sign 3×2 is read '3 by 2'.

3.3   Types of Matrices

1. A Square Matrix

A square matrix is a matrix with the number of rows and columns equal.

Examples of square matrices are $\begin{pmatrix} 0 & 1 & 2 \\ 8 & 3 & 7 \\ 5 & -4 & 6 \end{pmatrix}, \begin{pmatrix} 2 & 1 \\ 4 & 3 \end{pmatrix}$.

In a square matrix such as $\begin{pmatrix} 0 & 1 & 2 \\ 8 & 3 & 7 \\ 5 & -4 & 6 \end{pmatrix}$, the entries 0, 3 and 6 are in the **leading diagonal**. The entries 2, 3 and 5 are in the **minor diagonal**.

$\begin{pmatrix} 0 & 1 & 2 \\ 8 & 3 & 7 \\ 5 & -4 & 6 \end{pmatrix}$ ← Minor diagonal
                                                     ← Leading diagonal

2. A Diagonal Matrix

A diagonal matrix is a square matrix with all the elements zeros except those in the leading diagonal.

Examples of diagonal matrices are $\begin{pmatrix} 1 & 0 & 0 \\ 0 & 3 & 0 \\ 0 & 0 & 6 \end{pmatrix}, \begin{pmatrix} 2 & 0 \\ 0 & 3 \end{pmatrix}$.

Note that a diagonal matrix is necessarily a square matrix.

3. A Unit or Identity Matrix

The **unit** or **identity Matrix** usually denoted by **I** is a diagonal matrix with all the elements in the leading diagonal ones. The unit or identity Matrix is a very important matrix due to its applications.

Examples identity matrices are $\begin{pmatrix} 1 & 0 & 0 \\ 0 & 1 & 0 \\ 0 & 0 & 1 \end{pmatrix}, \begin{pmatrix} 1 & 0 \\ 0 & 1 \end{pmatrix}$.

4. A rectangular Matrix

A rectangular matrix is a matrix with the number of rows different from the number of columns.

Examples of rectangular matrices are $\begin{pmatrix} -1 & 3 \\ 0 & -4 \\ 1 & 5 \end{pmatrix}, \begin{pmatrix} 0 & 1 & 2 \\ 3 & 0 & 4 \end{pmatrix}$.

5. A Column Matrix

A column matrix is a matrix with all the elements in a single column.

Examples of column matrices are $\begin{pmatrix} 7 \\ 1 \\ -9 \end{pmatrix}, \begin{pmatrix} -2 \\ 4 \end{pmatrix}$.

6. A Row Matrix

A row matrix is a matrix with all the elements in a single row.

Examples of row matrices are $(7 \quad 1 \quad -9), (-2 \quad 4)$.

7. A Zero or Null Matrix

A zero or null matrix is one with all the elements zeros.

Module 10, Topic 3: Matrices

Examples of zero or null matrices are $\begin{pmatrix} 0 & 0 \\ 0 & 0 \end{pmatrix}, \begin{pmatrix} 0 & 0 & 0 \\ 0 & 0 & 0 \end{pmatrix}$.

We can classify some matrices in more than one group.

For instance, $\begin{pmatrix} 0 & 0 & 0 \\ 0 & 0 & 0 \end{pmatrix}$ is a rectangular zero matrix and $\begin{pmatrix} 0 & 0 \\ 0 & 0 \end{pmatrix}$ is a square zero matrix.

Exercise 3:1

1. State the sizes of the following matrices.

   (a) $\begin{pmatrix} 1 & 0 \\ 0 & 0 \end{pmatrix}$ (b) $\begin{pmatrix} 1 & 5 & 2 \\ 3 & 1 & 6 \end{pmatrix}$ (c) $\begin{pmatrix} -2 \\ 1 \\ -4 \end{pmatrix}$ (d) $\begin{pmatrix} 3 & 0 & 0 \\ 0 & -5 & 0 \\ 0 & 0 & -10 \end{pmatrix}$

   (e) $\begin{pmatrix} 0 & 0 \\ 0 & 0 \\ 0 & 0 \end{pmatrix}$ (f) (7) (g) $\begin{pmatrix} 2 & 1 & 4 \\ 0 & -1 & 0 \\ 1 & 3 & -5 \end{pmatrix}$ (h) $(8 \;\; 1 \;\; 4)$

   (i) $\begin{pmatrix} 1 & 2 \\ 3 & 0 \\ 7 & -3 \end{pmatrix}$ (j) $\begin{pmatrix} 1 & 0 & 0 \\ 0 & 1 & 0 \\ 0 & 0 & 1 \end{pmatrix}$

2. Classify the matrices in question (1) above.

3.4 Equivalent Matrices and Equal Matrices

Equivalent matrices are matrices which have exactly the same size or order. Two matrices **A** and **B** are equal if and only if:

(i) They are equivalent (have the same size or order) and.

(ii) Their corresponding elements are equal.

Let $\mathbf{A} = \begin{pmatrix} 2 & -1 & 1 \\ 0 & 5 & 3 \end{pmatrix}$ and $\mathbf{B} = \begin{pmatrix} u & v & w \\ x & y & z \end{pmatrix}$.

Then **A** and **B** are equivalent no matter the values of u, v, w, x, y, z because they have the same size 2×3. For **A** and **B** to be equal, $u = 2, v = -1, w = 1, x = 0, y = 5$ and $z = 3$.

Note that $\begin{pmatrix} 1 & 3 \\ 2 & 5 \end{pmatrix} \neq \begin{pmatrix} 3 & 1 \\ 2 & 5 \end{pmatrix} \neq \begin{pmatrix} 1 & 2 \\ 3 & 5 \end{pmatrix}$, because though equivalent and have the same elements, corresponding elements are not equal.

Example

Find x and y given that $\begin{pmatrix} 3x & y \\ 0 & 4x \end{pmatrix} = \begin{pmatrix} 6 & 2 \\ 0 & 8 \end{pmatrix}$.

Solution
Equating corresponding entries,
$3x = 6 \Rightarrow x = 2$ or $4x = 8 \Rightarrow x = 2$ and $y = 2$

Exercise 3:2

1. Find x and y given that

   (a) $\begin{pmatrix} 0 & 2y \\ 5x & -y \end{pmatrix} = \begin{pmatrix} 0 & 8 \\ -15 & -4 \end{pmatrix}$
   (b) $\begin{pmatrix} x+2 \\ y-1 \end{pmatrix} = \begin{pmatrix} 3 \\ -1 \end{pmatrix}$
   (c) $\begin{pmatrix} 2x \\ 3y \end{pmatrix} = \begin{pmatrix} 10 \\ 12 \end{pmatrix}$

   (d) $\begin{pmatrix} x+1 \\ x+y \end{pmatrix} = \begin{pmatrix} 4 \\ 1 \end{pmatrix}$
   (e) $\begin{pmatrix} 0+3y \\ 4x+y \end{pmatrix} = \begin{pmatrix} 6 \\ 2 \end{pmatrix}$

2. Find x, y and z in each of the following cases.

   (a) $\begin{pmatrix} 2x & y & 0 \\ x & -3y & 5z \end{pmatrix} = \begin{pmatrix} -6 & 4 & 0 \\ -3 & -12 & 2 \end{pmatrix}$

   (b) $\begin{pmatrix} 4x & 3y \\ -2x & 0 \\ 0 & 5z \end{pmatrix} = \begin{pmatrix} -8 & 12 \\ 4 & 0 \\ 0 & -10 \end{pmatrix}$

3. Find x and y in the following:

   (a) $\begin{pmatrix} x & 2y \\ 0 & -2 \end{pmatrix} = \begin{pmatrix} 1 & -8 \\ 0 & -2 \end{pmatrix}$
   (b) $\begin{pmatrix} x+3 \\ 2-y \end{pmatrix} = \begin{pmatrix} 1 \\ -3 \end{pmatrix}$

   (c) $\begin{pmatrix} 2x-y \\ x+y \end{pmatrix} = \begin{pmatrix} 3 \\ -9 \end{pmatrix}$
   (d) $\begin{pmatrix} x+2y \\ 2y-x \end{pmatrix} = \begin{pmatrix} 0 \\ 2x-y \end{pmatrix}$

   (e) $\begin{pmatrix} 7 & -6 \\ -15 & 3y \end{pmatrix} = \begin{pmatrix} 2x+y & 2y \\ -3x & -9 \end{pmatrix}$
   (f) $\begin{pmatrix} 3 & x+1 \\ y & 5 \end{pmatrix} = \begin{pmatrix} x-1 & 5 \\ x-3y & 5 \end{pmatrix}$

Module 10, Topic 3: Matrices

3.5 Addition and Subtraction of Matrices

For addition and subtraction of matrices to be possible, the matrices must be equivalent. If this condition is satisfied, the matrices are then said to be **compatible** for matrix addition or subtraction and corresponding elements can be added or subtracted as the case may be.

Example

Let $A = \begin{pmatrix} 1 & 3 \\ 1 & 2 \end{pmatrix}$ and $B = \begin{pmatrix} 0 & -1 \\ 2 & 3 \end{pmatrix}$.

Find (i) $A + B$ (ii) $A - B$

Solution

(i) $A + B = \begin{pmatrix} 1 & 3 \\ 1 & 2 \end{pmatrix} + \begin{pmatrix} 0 & -1 \\ 2 & 3 \end{pmatrix} = \begin{pmatrix} 1 & 2 \\ 3 & 5 \end{pmatrix}$

(ii) $A - B = \begin{pmatrix} 1 & 3 \\ 1 & 2 \end{pmatrix} - \begin{pmatrix} 0 & -1 \\ 2 & 3 \end{pmatrix} = \begin{pmatrix} 1 & 4 \\ -1 & -1 \end{pmatrix}$

Exercise 3:3

1. Evaluate the following.

   (a) $\begin{pmatrix} -7 \\ 3 \end{pmatrix} + \begin{pmatrix} 9 \\ -5 \end{pmatrix}$ (b) $\begin{pmatrix} 4 & 3 \\ 1 & 2 \end{pmatrix} + \begin{pmatrix} 5 & -2 \\ 2 & 6 \end{pmatrix}$

   (c) $\begin{pmatrix} -7 \\ 3 \end{pmatrix} - \begin{pmatrix} 9 \\ -5 \end{pmatrix}$ (d) $\begin{pmatrix} 4 & 3 \\ 1 & 2 \end{pmatrix} - \begin{pmatrix} 5 & -2 \\ 2 & 6 \end{pmatrix}$

2. Given that $E = \begin{pmatrix} 1 & 2 & 4 \\ 3 & -1 & 8 \\ 2 & 5 & -7 \end{pmatrix}$ and $F = \begin{pmatrix} 0 & 3 & -5 \\ 2 & -3 & 2 \\ 1 & 4 & 7 \end{pmatrix}$, find:

   (a) $E - F$ (b) $F - E$

3. Given that $A = \begin{pmatrix} 1 & 2 & 0 \\ 2 & 3 & 0 \end{pmatrix}$, $B = \begin{pmatrix} 0 & 1 & 2 \\ 0 & 2 & 3 \end{pmatrix}$ and $C = \begin{pmatrix} 1 & 0 & 2 \\ 3 & 0 & 4 \end{pmatrix}$. Find

   (a) $A + B$ (b) $B + A$ (c) $B + C$ (d) $A - B$ (e) $B - A$ (f) $B - C$
   (g) $(A + B) + C$ (h) $A + (B + C)$ (i) $(A - B) - C$ (j) $A - (B - C)$
   Say whether or not:
   (k) Addition of matrices is commutative.

(l) Subtraction of matrices is commutative.
(m) Addition of matrices is associative.
(n) Subtraction of matrices is associative.

4. Given that $\mathbf{A} = \begin{pmatrix} 3 & -1 \\ 2 & 5 \end{pmatrix}, \mathbf{B} = \begin{pmatrix} -3 & 1 \\ -2 & -5 \end{pmatrix}$.

   Find (a) $\mathbf{A} + \mathbf{B}$ (b) $\mathbf{B} + \mathbf{A}$ (c) Comment on your result in (a) and (b).

5. Given that $\mathbf{A} = \begin{pmatrix} 3 & -1 \\ 2 & 5 \end{pmatrix}$ and $\varnothing = \begin{pmatrix} 0 & 0 \\ 0 & 0 \end{pmatrix}$.

   Find (a) $\mathbf{A} + \varnothing$ (b) $\varnothing + \mathbf{A}$
   (c) Suggest a name for $\varnothing$ with respect to its relationship to $\mathbf{A}$.

6. Find a, b, c, d and e, given that
$$\begin{pmatrix} 3 & 2 & 5 \\ 1 & 4 & c \end{pmatrix} + \begin{pmatrix} a & b & c \\ 0 & a & b \end{pmatrix} = \begin{pmatrix} 4 & a & b \\ a & d & e \end{pmatrix}$$

7. Find the unknowns in each of the following:

   (a) $\begin{pmatrix} 2 & 4 \\ 3 & z \end{pmatrix} + \begin{pmatrix} x & y \\ 3 & 4 \end{pmatrix} = \begin{pmatrix} 4 & 4 \\ w & 0 \end{pmatrix}$

   (b) $\begin{pmatrix} x & y \\ z & w \end{pmatrix} - \begin{pmatrix} 2 & 1 \\ 5 & 3 \end{pmatrix} = \begin{pmatrix} 1 & 0 \\ 0 & 1 \end{pmatrix}$

3.6 Scalar Multiplication of Matrices

To multiply a matrix by a scalar, simply multiply each element of the matrix by the scalar.

 Example

Evaluate the following:

(a) $2\begin{pmatrix} 2 & 3 \\ 1 & 4 \end{pmatrix}$ (b) $-3\begin{pmatrix} 4 & 1 & 0 \\ 5 & 2 & 7 \end{pmatrix}$ (c) $\frac{1}{3}\begin{pmatrix} 6 & 9 \\ -15 & 12 \end{pmatrix}$

Solution

(a) $2\begin{pmatrix} 2 & 3 \\ 1 & 4 \end{pmatrix} = \begin{pmatrix} 2\times 2 & 2\times 3 \\ 2\times 1 & 2\times 4 \end{pmatrix} = \begin{pmatrix} 4 & 6 \\ 2 & 8 \end{pmatrix}$

(b) $-3\begin{pmatrix} 4 & 1 & 0 \\ 5 & 2 & 7 \end{pmatrix} = \begin{pmatrix} -3\times 4 & -3\times 1 & -3\times 0 \\ -3\times 5 & -3\times 2 & -3\times 7 \end{pmatrix} = \begin{pmatrix} -12 & -3 & 0 \\ -15 & -6 & -21 \end{pmatrix}$

(c) $\dfrac{1}{3}\begin{pmatrix} 6 & 9 \\ -15 & 12 \end{pmatrix} = \begin{pmatrix} \dfrac{1}{3} \times 6 & \dfrac{1}{3} \times 9 \\ \dfrac{1}{3} \times -15 & \dfrac{1}{3} \times 12 \end{pmatrix} = \begin{pmatrix} 2 & 3 \\ -5 & 4 \end{pmatrix}$

Exercise 3:4

(1) Evaluate $-\dfrac{1}{2}\begin{pmatrix} -4 & 0 & -1 \\ 3 & 1 & -2 \\ 0 & 8 & 5 \end{pmatrix}$.

(2) Given that $x\begin{pmatrix} 1 \\ 2 \end{pmatrix} + 3\begin{pmatrix} 9 \\ -4 \end{pmatrix} = \begin{pmatrix} 30 \\ -6 \end{pmatrix}$, find x.

(3) Find x and y given that $x\begin{pmatrix} 1 \\ 0 \end{pmatrix} - y\begin{pmatrix} 0 \\ 4 \end{pmatrix} = \begin{pmatrix} 5 \\ 8 \end{pmatrix}$.

(4) Find the unknown if:

(i) $2\begin{pmatrix} x \\ y \end{pmatrix} = \begin{pmatrix} -4 \\ -10 \end{pmatrix}$

(ii) $\begin{pmatrix} x \\ y \end{pmatrix} = -\dfrac{1}{3}\begin{pmatrix} -27 \\ 21 \end{pmatrix}$

(iii) $a\begin{pmatrix} 0 \\ 3 \end{pmatrix} + b\begin{pmatrix} 1 \\ 0 \end{pmatrix} = \begin{pmatrix} 11 \\ 6 \end{pmatrix}$

3.7 Multiplication of Matrices

If **A** and **B** are matrices, then the product **A** × **B** read "A cross B" is possible only if the number of columns in the matrix **A** are equal to the number of rows in the matrix **B**. This means that if the size of the matrix **A** is $m \times n$, the size of **B**, must be $n \times p$. If this condition is satisfied, we say **A** and **B** are **compatible** or **conformable** for matrix multiplication and the product **A** × **B** exists. The size of **A** × **B** will then be $m \times p$

Therefore, (m by n) × (n by p) = m by p

We can simply **A** × **B** as **AB**.

Example

1. Compute the following.

   (a) $(a \quad b)\begin{pmatrix} x \\ y \end{pmatrix}$ (b) $(a \quad b)\begin{pmatrix} w & x \\ y & z \end{pmatrix}$ (c) $(4 \quad 5)\begin{pmatrix} 2 \\ 3 \end{pmatrix}$

   (d) $(2 \quad 4)\begin{pmatrix} 1 & 4 \\ 2 & 0 \end{pmatrix}$ (e) $\begin{pmatrix} a & b \\ c & d \end{pmatrix}\begin{pmatrix} w & x \\ y & z \end{pmatrix}$

   (f) $\begin{pmatrix} 1 & 0 \\ 2 & 3 \end{pmatrix}\begin{pmatrix} -3 & 2 \\ 1 & 7 \end{pmatrix}$ (g) $\begin{pmatrix} 2 & 7 \\ 1 & 3 \end{pmatrix}\begin{pmatrix} 1 & 4 & -3 \\ 5 & 2 & 0 \end{pmatrix}$

   Solution

   (a) $(a \quad b)\begin{pmatrix} x \\ y \end{pmatrix} = (ax + by)$

   (b) $(a \quad b)\begin{pmatrix} w & x \\ y & z \end{pmatrix} = (aw + by \quad ax + bz)$

   (c) $(4 \quad 5)\begin{pmatrix} 2 \\ 3 \end{pmatrix} = (4(2) + 5(3)) = (8 + 15) = (23)$

   (d) $(2 \quad 4)\begin{pmatrix} 1 & 4 \\ 2 & 0 \end{pmatrix} = (2(1) + 4(2) \quad 2(4) + 4(0)) = (10 \quad 8)$

   (e) $\begin{pmatrix} a & b \\ c & d \end{pmatrix}\begin{pmatrix} w & x \\ y & z \end{pmatrix} = \begin{pmatrix} aw + by & ax + bz \\ cw + dy & cx + dz \end{pmatrix}$

   (f) $\begin{pmatrix} 1 & 0 \\ 2 & 3 \end{pmatrix}\begin{pmatrix} -3 & 2 \\ 1 & 7 \end{pmatrix} = \begin{pmatrix} 1(-3) + 0(1) & 1(2) + 0(7) \\ 2(-3) + 3(1) & 2(2) + 3(7) \end{pmatrix} = \begin{pmatrix} -3 & 2 \\ -3 & 25 \end{pmatrix}$

   (g) $\begin{pmatrix} 2 & 7 \\ 1 & 3 \end{pmatrix}\begin{pmatrix} 1 & 4 & -3 \\ 5 & 4 & 0 \end{pmatrix} = \begin{pmatrix} 2+35 & 8+14 & -6+0 \\ 1+15 & 4+6 & -3+0 \end{pmatrix} = \begin{pmatrix} 37 & 22 & -6 \\ 16 & 10 & -3 \end{pmatrix}$

2. Let $\mathbf{A} = \begin{pmatrix} 1 & 2 \\ 1 & 4 \end{pmatrix}$ and $\mathbf{B} = \begin{pmatrix} 4 & 1 \\ 3 & -2 \end{pmatrix}$. Find (i) **AB** (ii) **BA**

   Solution

   (i) $\mathbf{AB} = \begin{pmatrix} 1 & 2 \\ 1 & 4 \end{pmatrix}\begin{pmatrix} 4 & 1 \\ 3 & -2 \end{pmatrix} = \begin{pmatrix} 10 & -3 \\ 16 & -7 \end{pmatrix}$

(ii) $\mathbf{BA} = \begin{pmatrix} 4 & 1 \\ 3 & -2 \end{pmatrix} \begin{pmatrix} 1 & 2 \\ 1 & 4 \end{pmatrix} = \begin{pmatrix} 5 & 12 \\ 1 & -2 \end{pmatrix}$

3. Given that $\mathbf{P} = \begin{pmatrix} 3 & 4 \\ 5 & 6 \end{pmatrix}$ and $\mathbf{Q} = \begin{pmatrix} 2 \\ 1 \end{pmatrix}$, find **PQ**.

Solution

$\mathbf{PQ} = \begin{pmatrix} 3 & 4 \\ 5 & 6 \end{pmatrix} \begin{pmatrix} 2 \\ 1 \end{pmatrix} = \begin{pmatrix} 10 \\ 16 \end{pmatrix}$

4. Find **AB** given that $\mathbf{A} = \begin{pmatrix} 0 & -1 \\ 3 & -2 \end{pmatrix}$ and $\mathbf{B} = \begin{pmatrix} 2 & 4 & 3 \\ 1 & 5 & 0 \end{pmatrix}$.

Solution

$\mathbf{AB} = \begin{pmatrix} 0 & -1 \\ 3 & -2 \end{pmatrix} \begin{pmatrix} 2 & 4 & 3 \\ 1 & 5 & 0 \end{pmatrix} = \begin{pmatrix} -1 & -5 & 0 \\ 4 & 2 & 9 \end{pmatrix}$

In the product **AB**, **A** is pre-multiplied by **B**, while **B** is said to be post-multiplied by **A**.

Notice that $\mathbf{AB} \neq \mathbf{BA}$. Therefore, matrix multiplication is not commutative.

Exercise 3:5

1. Let $\mathbf{A} = \begin{pmatrix} 2 & -3 \\ 4 & 2 \end{pmatrix}$ and $\mathbf{B} = \begin{pmatrix} -1 & 0 \\ 2 & -3 \end{pmatrix}$. Find **AB**.

2. $\mathbf{C} = \begin{pmatrix} 5 & 7 \end{pmatrix}$ and $\mathbf{D} = \begin{pmatrix} 3 & -7 & 2 \\ -4 & 0 & 1 \end{pmatrix}$. Evaluate **CD**.

3. Given that $\mathbf{A} = \begin{pmatrix} 3 & 7 \\ 5 & -2 \end{pmatrix}$ and $\mathbf{D} = \begin{pmatrix} 1 & 0 \\ 0 & 1 \end{pmatrix}$, find
    (i) **A×D**   (ii) **D×A**   (iii) $\mathbf{A}^2$

4. Evaluate the following and comment on your results.
    (a) $\begin{pmatrix} 1 & 3 \\ 5 & 4 \end{pmatrix} \begin{pmatrix} 1 & 0 \\ 0 & 1 \end{pmatrix}$ (b) $\begin{pmatrix} 1 & 0 \\ 0 & 1 \end{pmatrix} \begin{pmatrix} 1 & 3 \\ 5 & 4 \end{pmatrix}$ (c) $\begin{pmatrix} 1 & 3 \\ 5 & 4 \end{pmatrix} \begin{pmatrix} 0 & 0 \\ 0 & 0 \end{pmatrix}$

(d) $\begin{pmatrix} 0 & 0 \\ 0 & 0 \end{pmatrix}\begin{pmatrix} 1 & 3 \\ 5 & 4 \end{pmatrix}$ (e) $\begin{pmatrix} 1 & 2 \\ 5 & 4 \\ 2 & 6 \end{pmatrix}\begin{pmatrix} 1 & 0 \\ 0 & 1 \end{pmatrix}$ (f) $\begin{pmatrix} 1 & 2 \\ 5 & 4 \\ 2 & 6 \end{pmatrix}\begin{pmatrix} 0 & 0 \\ 0 & 0 \end{pmatrix}$

5. Let $A = \begin{pmatrix} 2 & -4 \\ 1 & 3 \end{pmatrix}$, $B = \begin{pmatrix} 2 & 3 \\ 1 & 4 \end{pmatrix}$ and $C = \begin{pmatrix} 0 & 2 \\ -1 & 3 \end{pmatrix}$

   (i) Evaluate the following:
   - (a) A^2
   - (b) B^2
   - (c) AB
   - (d) BA
   - (e) BC
   - (f) $(AB)C$
   - (g) $A(BC)$
   - (h) $A + B$
   - (i) $A - B$
   - (j) $(A + B)^2$
   - (k) $(A - B)^2$
   - (l) $A^2 - B^2$
   - (m) $(A + B)(A - B)$
   - (n) $A^2 - 2AB + B^2$
   - (o) $A^2 + 2AB + B^2$

   (ii) State whether the following operations are true of matrix multiplication.
   - (a) $AB = BA$
   - (b) $(AB)C = A(BC)$
   - (c) $(A + B)^2 = A^2 + 2AB + B^2$
   - (d) $(A - B)^2 = A^2 - 2AB + B^2$
   - (e) $(A + B)(A - B) = A^2 - B^2$

6. Compute the following products:

   (a) $\begin{pmatrix} 5 & -3 \\ -4 & 1 \end{pmatrix}\begin{pmatrix} x \\ y \end{pmatrix}$ (b) $\begin{pmatrix} 3 & 4 \\ 5 & 1 \end{pmatrix}\begin{pmatrix} a \\ b \end{pmatrix}$ (c) $\begin{pmatrix} 2 & 1 \\ -2 & 3 \end{pmatrix}\begin{pmatrix} r \\ s \end{pmatrix}$

   (d) $\begin{pmatrix} 4 & 1 \\ 2 & 0 \end{pmatrix}\begin{pmatrix} p \\ q \end{pmatrix}$ (e) $\begin{pmatrix} 4 & 2 \\ 3 & 5 \end{pmatrix}\begin{pmatrix} x \\ y \end{pmatrix}$ (f) $\begin{pmatrix} -2 & -3 \\ -2 & 1 \end{pmatrix}\begin{pmatrix} x \\ y \end{pmatrix}$

3.8 The Transpose of a Matrix

We obtain the transpose A^T of a matrix A by interchanging the rows and columns of the matrix.

Example

Find the transpose of **A** and **B** given that:

(a) $A = \begin{pmatrix} 1 & 4 \\ 2 & 5 \end{pmatrix}$ (b) $B = \begin{pmatrix} 1 & 3 \\ 2 & 4 \\ 7 & -9 \end{pmatrix}$

Solution

(a) $A^T = \begin{pmatrix} 1 & 2 \\ 4 & 5 \end{pmatrix}$ (b) $B^T = \begin{pmatrix} 1 & 2 & 7 \\ 3 & 4 & -9 \end{pmatrix}$

Module 10, Topic 3: Matrices

 Exercise 3:6

Write down the transpose of each of the following matrices.

(a) $\begin{pmatrix} 1 & 3 \\ 0 & 4 \end{pmatrix}$ (b) $\begin{pmatrix} 1 & 2 & 7 \end{pmatrix}$ (c) $\begin{pmatrix} -1 \\ 0 \\ 1 \end{pmatrix}$ (d) $\begin{pmatrix} 2 & 0 & -4 \\ 1 & 8 & 6 \\ -3 & 5 & 1 \end{pmatrix}$

3.9   The Determinant of a 2 × 2 Matrix

We obtain the **determinant** of a 2×2 matrix $A = \begin{pmatrix} a & b \\ c & d \end{pmatrix}$, denoted by det **A**, $\det\begin{pmatrix} a & b \\ c & d \end{pmatrix}$, $|A|$ or $\begin{vmatrix} a & b \\ c & d \end{vmatrix}$ and by finding the difference between the product of the leading diagonal elements and the minor diagonal elements. Hence,

$$\text{Det } A = ad - bc$$

Note! $\begin{pmatrix} a & b \\ c & d \end{pmatrix} \neq \begin{vmatrix} a & b \\ c & d \end{vmatrix}$

 Example

Evaluate (a) $\begin{vmatrix} 3 & 2 \\ 1 & 5 \end{vmatrix}$ (b) $\begin{vmatrix} 4 & 2 \\ -1 & 3 \end{vmatrix}$

Solution

(a) $\begin{vmatrix} 3 & 2 \\ 1 & 5 \end{vmatrix} = 3(5) - 2(1) = 13$ (b) $\begin{vmatrix} 4 & 2 \\ -1 & 3 \end{vmatrix} = 4(3) - 2(-1) = 14$

3.10 Singular Matrices

A **singular matrix** is a matrix whose determinant is equal to zero.

Example

Given that $\mathbf{A} = \begin{pmatrix} 2 & -1 \\ 4 & -2 \end{pmatrix}$ and $\mathbf{B} = \begin{pmatrix} 1 & 1 \\ 1 & 1 \end{pmatrix}$, find (i) det $\mathbf{A}$ (ii) $|\mathbf{B}|$.

Solution
(i) det $\mathbf{A} = 2(-2) - 4(-1) = 0$ (ii) $|\mathbf{B}| = 1(1) - 1(1) = 0$.
Therefore, both $\mathbf{A}$ and $\mathbf{B}$ are singular matrices.

Exercise 3:7

1. Find the determinant of $\mathbf{A} = \begin{pmatrix} 3 & 4 \\ 5 & 6 \end{pmatrix}$.

2. Find the value of a for which the matrix $\begin{pmatrix} a & 3-a \\ 2 & 1 \end{pmatrix}$ is singular.

3. Given that the matrix $\begin{pmatrix} a & 1 \\ 4 & a \end{pmatrix}$ is singular, find the possible values of a.

4. Which of the following matrices is/are singular?
$\begin{pmatrix} -2 & 2 \\ 2 & 2 \end{pmatrix}, \begin{pmatrix} -2 & -2 \\ 2 & 2 \end{pmatrix}, \begin{pmatrix} -2 & -2 \\ 2 & -2 \end{pmatrix}, \begin{pmatrix} -2 & -2 \\ -2 & 2 \end{pmatrix}, \begin{pmatrix} 2 & -2 \\ -2 & -2 \end{pmatrix}$.

5. Find $\begin{pmatrix} 2 & -3 \\ -2 & 4 \end{pmatrix} \begin{pmatrix} -2 & 1 \\ -5 & 3 \end{pmatrix}$ and calculate its determinant.

Multiple Choice Exercise 3

1. The order of the matrices $\begin{pmatrix} 3 & 1 & 2 \\ 4 & 0 & 3 \end{pmatrix}$ and $\begin{pmatrix} 4 \\ 0 \\ 3 \end{pmatrix}$ are respectively:

   [A] 2×3 and 1×3   [B] 2×3 and 3×1   [C] 3×2 and 1×3   [D] 3×2 and 3×1

2. **A** is a 2 by 3 matrix and **B** is a 4 by 2 matrix. The number of elements in **A** and **B** are respectively:
   [A] 4 and 6   [B] 6 and 4   [C] 8 and 6   [D] 6 and 8

3. An example of a 3 by 2 matrix is:

   [A] $\begin{pmatrix} 3 \\ 2 \end{pmatrix}$ [B] $\begin{pmatrix} 0 & 2 & 4 \\ 0 & 0 & 1 \end{pmatrix}$ [C] $\begin{pmatrix} 0 & 1 \\ 0 & -1 \\ 0 & 4 \end{pmatrix}$ [D] $\begin{pmatrix} 3 & 0 \\ 0 & 3 \end{pmatrix}$

4. Let $\mathbf{A} = \begin{pmatrix} 4 & 1 \\ -2 & 3 \end{pmatrix}$. The transpose of **A** is:

   [A] $\begin{pmatrix} 4 & -2 \\ 1 & 3 \end{pmatrix}$ [B] $\begin{pmatrix} 3 & -2 \\ 1 & 4 \end{pmatrix}$ [C] $\begin{pmatrix} 3 & 1 \\ -2 & 4 \end{pmatrix}$ [D] $\begin{pmatrix} 4 & 1 \\ -2 & 3 \end{pmatrix}$

5. Given that **A** is a 3 by 2 matrix and **B** is a 2 by 4 matrix. The matrix product **AB** will have size:
   [A] 2 by 3   [B] 3 by 4   [C] 3 by 2   [D] 2 by 4

6. Given that $\mathbf{A} = \begin{pmatrix} 2 & -3 \\ -2 & 4 \end{pmatrix}$ and $\mathbf{B} = \begin{pmatrix} -2 & 1 \\ -5 & 3 \end{pmatrix}$. The matrix product **AB** is:

   [A] $\begin{pmatrix} -11 & 7 \\ 16 & 10 \end{pmatrix}$ [B] $\begin{pmatrix} 11 & -7 \\ 16 & 10 \end{pmatrix}$ [C] $\begin{pmatrix} 11 & -7 \\ -16 & 10 \end{pmatrix}$ [D] $\begin{pmatrix} 11 & 7 \\ -16 & 10 \end{pmatrix}$

7. Given that $\mathbf{A} = \begin{pmatrix} 2 & -3 \\ -2 & 4 \end{pmatrix}$ and $\mathbf{B} = \begin{pmatrix} -2 & 1 \\ -5 & 3 \end{pmatrix}$. The determinant of the product **AB** is:
   [A] 2   [B] 222   [C] −222   [D] −2

8. Given that $\mathbf{A} = \begin{pmatrix} 0 & 2 \\ 4 & -2 \\ -1 & 1 \end{pmatrix}$ and $\mathbf{B} = \begin{pmatrix} -2 & 4 & -1 & 2 \\ 0 & 3 & 0 & 1 \end{pmatrix}$. The size of the matrix product **AB** is:
   [A] 2 × 3   [B] 3 × 4   [C] 3 × 2   [D] 2 × 4

9. The singular matrix among the following matrices is:

   [A] $\begin{pmatrix} -2 & 2 \\ 2 & 2 \end{pmatrix}$ [B] $\begin{pmatrix} -2 & -2 \\ 2 & 2 \end{pmatrix}$ [C] $\begin{pmatrix} -2 & -2 \\ 2 & -2 \end{pmatrix}$ [D] $\begin{pmatrix} -2 & -2 \\ -2 & 2 \end{pmatrix}$

10. The value of a for which the matrix $\begin{pmatrix} a & 3-a \\ 2 & 1 \end{pmatrix}$ is singular is:

[A] 2   [B] −2   [C] 3   [D] −3

11. The transpose of the matrix $\begin{pmatrix} 7 & 3 \\ -1 & 5 \end{pmatrix}$ is:

[A] $\begin{pmatrix} 5 & -3 \\ 1 & 7 \end{pmatrix}$ [B] $\begin{pmatrix} 7 & -1 \\ 3 & 5 \end{pmatrix}$ [C] $\begin{pmatrix} -7 & -1 \\ 3 & -5 \end{pmatrix}$ [D] $\begin{pmatrix} -7 & 1 \\ -3 & -5 \end{pmatrix}$

12. Given that $\begin{pmatrix} 2 & 1 \\ 1 & -1 \end{pmatrix}\begin{pmatrix} 1 & -2 \\ 1 & 2 \end{pmatrix} = 3\mathbf{A}$. The matrix **A** is:

[A] $\begin{pmatrix} 3 & -2 \\ 0 & -4 \end{pmatrix}$ [B] $\begin{pmatrix} 1 & \frac{2}{3} \\ 0 & \frac{4}{3} \end{pmatrix}$ [C] $\begin{pmatrix} 3 & \frac{2}{3} \\ 0 & \frac{4}{3} \end{pmatrix}$ [D] $\begin{pmatrix} 3 & -\frac{2}{3} \\ 0 & -\frac{4}{3} \end{pmatrix}$

13. The determinant of the matrix $\mathbf{A} = \begin{pmatrix} x & 7 \\ 4 & 2 \end{pmatrix}$ is 6. The value of x is:

[A] −17   [B] −34   [C] 17   [D] 34

14. Let $\mathbf{M} = \begin{pmatrix} 3 & 1 \\ 1 & 2 \end{pmatrix}$ and $\mathbf{N} = \begin{pmatrix} 1 & 4 \\ -3 & 2 \end{pmatrix}$. As a single matrix, $\mathbf{M} + 3\mathbf{N}$ is equal to:

15. [A] $\begin{pmatrix} 10 & 7 \\ 1 & 4 \end{pmatrix}$ [B] $\begin{pmatrix} 10 & 7 \\ -1 & 4 \end{pmatrix}$ [C] $\begin{pmatrix} 6 & 13 \\ 8 & -8 \end{pmatrix}$ [D] $\begin{pmatrix} 6 & 13 \\ -8 & 8 \end{pmatrix}$

16. The incorrect statement among the following is:
   [A] A square matrix is necessarily a diagonal matrix.
   [B] A diagonal matrix is necessarily a square matrix.
   [C] A unit matrix is necessarily a square matrix.
   [D] A unit matrix is necessarily a diagonal matrix.

Module 11

Plane Geometry

Family of Situations
Module 11 is an extension of module 2 and 6. At the end of the module; the student is expected to acquire many more competencies within the **families of situations** *'Representation and transformation of points and Plane Shapes within the Environment'*.

Categories of Action
The categories of action for module 11 include:
1. Interpretation and organization of the physical environment,
2. Production of plane shapes,
3. Transformation of the physical environment,
4. Determination of measures and position within the physical environment.

Credit
The module is expected to be covered within 6 weeks teaching 4 periods of 50 minutes per week (or within 24 periods).

Topic 4

CONGRUENCY AND SIMILARITY

Objectives

At the end of this topic, the learner should be able to:

1. State and use the conditions for two triangles to be congruent or similar.
2. Apply the knowledge of congruency and similarity to real life situations.
3. Identify congruent figures and similar figures in the environment.
4. Construct similar figures.
5. Compare areas of similar figures.
6. State and use Thales property.
7. Use ratio of corresponding sides to find the area of similar plane figures.
8. Apply the notion of scale factor to real life situations.

Module 11, Topic 4: Congruency and Similarity

4.1 Congruent Figures

Congruent figures are figures that have the same shape and the same size. Congruent plane figures are figures, which can fit on each other. The statement 'A is congruent to B' is written $A \equiv B$.

4.2 Congruent Triangles

Conditions for Triangles to be Congruent

1. If the three sides of one triangle are equal respectively to the three sides of another triangle then the triangles are congruent by **side-side-side** abbreviated **SSS**.

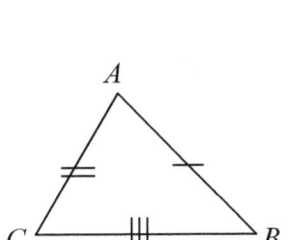
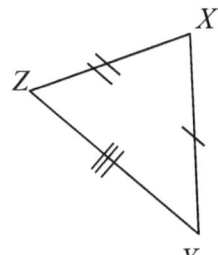

$AB = XY$
$BC = YZ$
$AC = XZ$

2. If two sides and the included angle of one triangle are equal respectively to two sides and the included angle of another triangle then the triangles are congruent by **side - angle-side** abbreviated **SAS**.

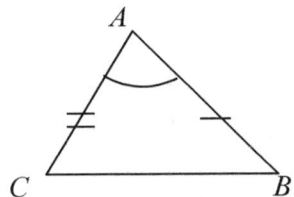
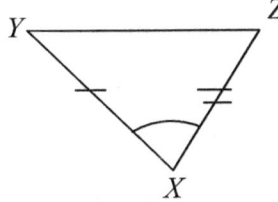

$AB = XY$
$AC = XZ$
$\angle A = \angle X$

3. If two angles and the included side of one triangle are equal respectively to two angles and the included side of another triangle then the triangles are congruent by **angle-side-angle** abbreviated **ASA**.

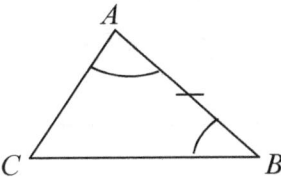
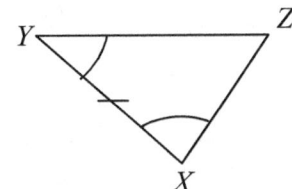

$\angle A = \angle Y$
$\angle C = \angle Z$
$AC = YZ$

4. If the hypotenuse and one side of a right-angled triangle are equal to the hypotenuse and one side of another right-angled triangle then the triangles are congruent by **right angle-hypotenuse- side** abbreviated **RHS**.

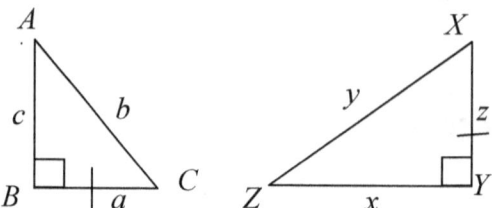

$\angle ABC = \angle XYZ = 90°$
$AC = XZ$
$AB = XY$

 Example

In the following figure, $PR = PS$ and Q and T are the mid-points of PR and PS respectively. Determine the pairs of triangles, which are congruent giving arguments and reasons leading to your answer.

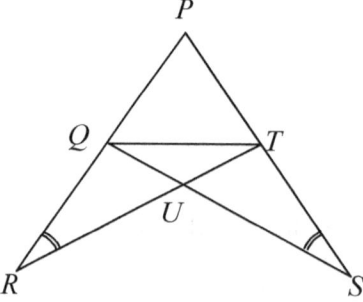

Solution

Argument	Reason
$PS = PR$	Given
$PQ = PT$	P and Q are mid-points of PR and PS
$\angle RPS$ is common to $\triangle PRT$ and $\triangle PSQ$	Shown on diagram
$\therefore \triangle PRT \equiv \triangle PSQ$	SAS
$RT = SQ$	$\triangle PRT = \triangle PSQ$ as proven above
$RQ = ST$	P and Q are mid-points of PR and PS
QT is common to $\triangle QRT$ and $\triangle TSQ$	Shown on diagram
$\therefore \triangle QRT \equiv \triangle TSQ$	SSS

Module 11, Topic 4: Congruency and Similarity

∠QUR = ∠TUS	Vert. opp. angles
∠QRT = ∠TSQ	△QRT=△TSQ
∠RQU = ∠RTS	Sum of ∠s of △
∴ △QRU ≡ △TSU	ASA

 Exercise 4:1

In problems 1 to 7, determine which pair of triangles, are congruent giving arguments leading to your answer.

1. In figure (a) below PQ is parallel to SR.

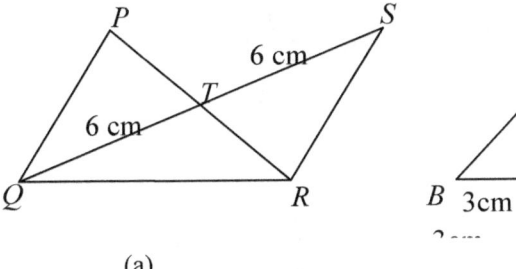

(a) (b)

2. In figure (b) above triangle ABE is isosceles $BC = DE = 3$ cm.

3. In figure (a) below, O is such that $OY = OZ$ and $\angle YOZ = \angle ZOX = \angle XOY = 120°$.

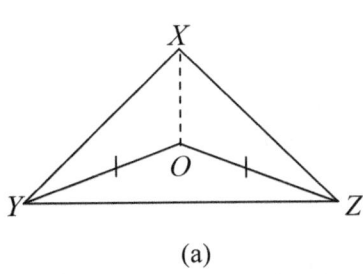 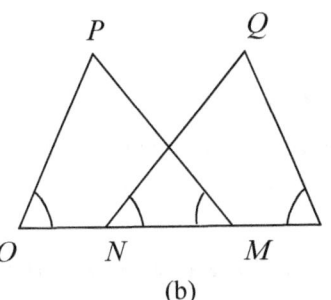

(a) (b)

4. In figure (b) above, $OM = NL$, $\angle OMP = \angle LNQ$ and $\angle POM = \angle QLN$.

5. In figure (a) below, PQRS is a parallelogram.

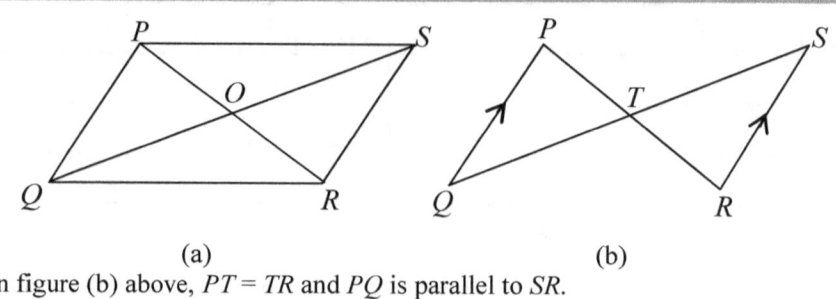

(a) (b)

6. In figure (b) above, $PT = TR$ and PQ is parallel to SR.
7. In figure (a) below,, $XO = WO$ and XW is parallel to YZ.

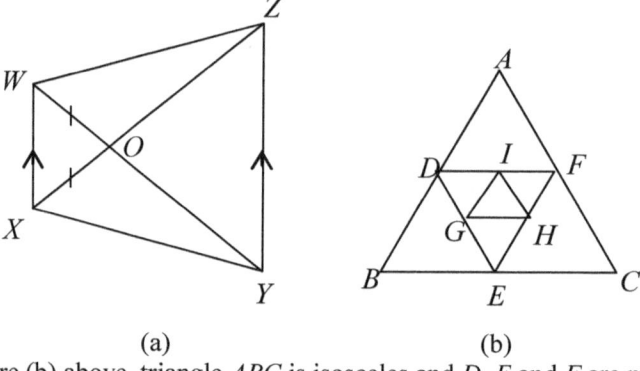

(a) (b)

8. In figure (b) above, triangle ABC is isosceles and D, E and F are mid-points of AB, BC and AC respectively. How many sets of congruent triangles are there in the figure? List each of these sets.

4.3 Similar Figures

Similar figures are figures that have the same shape but not necessarily the same size. The statement 'A is similar to B' is written $A///B$.

4.4 Similar Triangles

Similar triangles are triangles whose corresponding angles are congruent and whose corresponding sides are proportional.

Conditions for Triangles to be Similar

1. If two triangles are equiangular, then they are similar by **angle-angle-angle** abbreviated **AAA**.

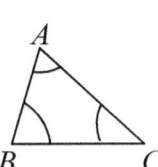 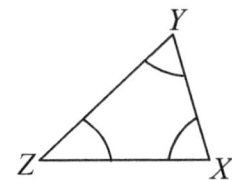

$\angle A = \angle X$
$\angle B = \angle Y$
$\angle C = \angle Z$

2. If the corresponding sides of two triangles are in a common ratio then they are similar by **side- side-side** abbreviated **SSS**.

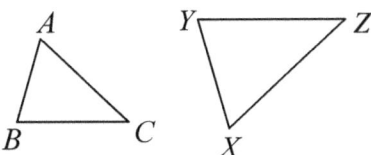

$$\frac{AB}{XY} = \frac{AC}{XZ} = \frac{BC}{YZ}$$

3. If an angle of one triangle is equal to an angle of another triangle and the sides containing this angle are in a common ratio then they are similar by **side-angle-side** abbreviated **SAS**.

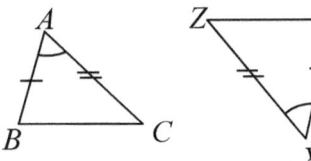

$\angle A = \angle X$
$$\frac{AB}{XY} = \frac{AC}{XZ}$$

 Example

1. Given that in Figure 32:17, $\triangle ABC \sim \triangle XYZ$. Find the value of x.

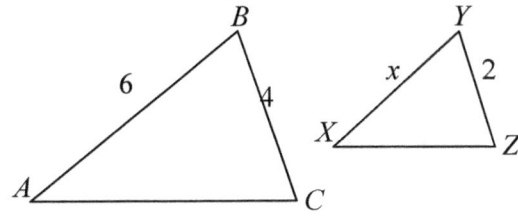

Solution

$$\frac{XY}{AB} = \frac{YZ}{BC}$$

$$\frac{x}{6} = \frac{2}{4}$$
$$x = 3$$

2. In Figure 33:18, $PQ \parallel ST$ calculate the value of y.

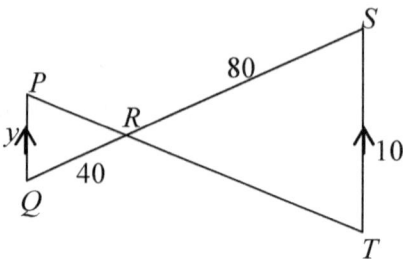

Solution

Since $PQ \parallel ST$, $\triangle RST \sim \triangle RQP \Rightarrow \frac{PQ}{TS} = \frac{RQ}{RS}$

$$\frac{y}{100} = \frac{40}{80}$$
$$\Rightarrow y = 50$$

 Exercise 4:2

1. The following figure shows a right-angled triangle CAD with BE parallel to CD. Given that $BE = 10$ cm, $AE = 6$ cm and $ED = 12$ cm, find
   (i) AB (ii) CD.

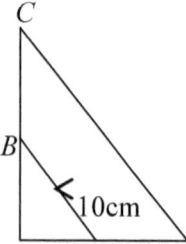

2. Determine which of the following sets consist entirely of elements, which are similar figures:
   (a) {triangles}   (b) {squares}   (c) {rectangles}
   (d) {parallelograms}   (e) {rhombuses}   (f) {trapeziums}
   (g) {hexagons}   (h) {circles}

3. In the following figure, $AB = 10$ cm, $PB = 3$ cm, $BC = 20$ cm and $AN = 9$ cm. AN is perpendicular to BC. If PQ is parallel to BC, calculate
   (i) the length of PQ (ii) the area of triangle APQ.

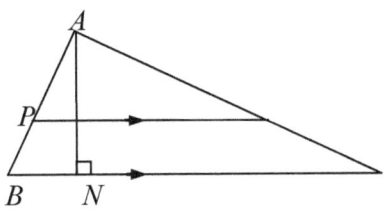

4. In the figure below the lengths of EC and AD are 8 cm and 6 cm respectively, ∠DAE = 60° and ∠AED = ∠ACB = 90°. Calculate the length of BC in cm to one decimal place.

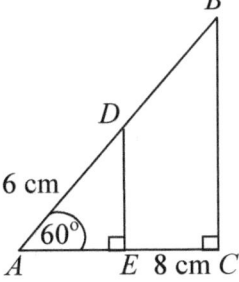

5. In the following figure, triangle XYZ of sides 9, 15 and n cm is an enlargement of triangle PQR of sides' m, 5 and 4.5 cm. Evaluate m and n.

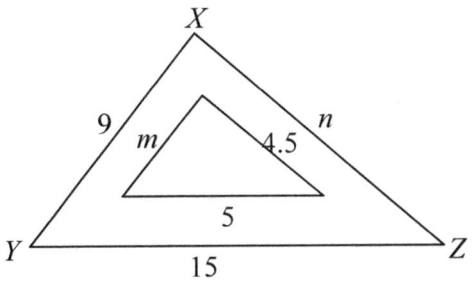

4.5 Ratio of Areas of Similar Figures

Let rectangle PQRS be an enlargement of rectangle ABCD with scale factor k. k is called the constant of proportionality.

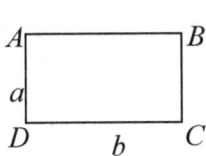
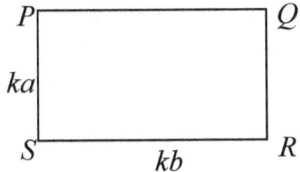

Area of ABCD = ab

Area of $PQRS = (ka)(kb) = k^2 ab$

$$\frac{\text{Area of } PQRS}{\text{Area of } ABCD} = \frac{k^2 ab}{ab} = k^2$$

$$\Rightarrow \frac{\text{Area of } PQRS}{\text{Area of } ABCD} = k^2$$

Therefore if the ratio of corresponding sides of two similar plane figures is $m:n = k$, then the ratio of their areas is $m^2:n^2 = k^2$.

 Example

The ratio of the corresponding sides of two similar triangles is 1:3. Calculate the area of the larger triangle given that the area of the smaller triangle is 8 cm².

Solution

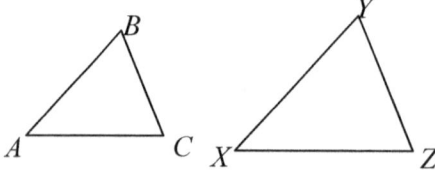

$$\frac{\text{Area of } \triangle XYZ}{\text{Area of } \triangle ABC} = \frac{3^2}{1^2}$$

Area of $\triangle XYZ = 9 \times$ Area $\triangle ABC = 9 \times 8 = 72$ cm²

 Exercise 4:3

1. On a map in which 1 cm represents 2 km, a plot of land is represented by a square of length 2.5 cm. Calculate the actual area, in km², of the plot of land.
2. The sides of a triangle are 5 cm, 6 cm, and 7 cm. The longest side of a similar triangle is 28 cm. Find the lengths of the other sides and the ratio of the area of the smaller triangle to the larger one.
3. The sides of a triangle are 6 cm, 9 cm, and 15 cm. The shortest side of a similar triangle is 2 cm. Find the lengths of the other sides and the ratio of the area of the smaller triangle to the larger one.
4. Given that the two trapeziums in the figure below are similar, find the value of x.

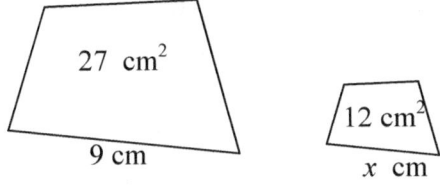

5. XY is parallel to BC and $\dfrac{AB}{AX} = \dfrac{3}{2}$. If the area of $\triangle AXY = 4$ cm², find the area of $\triangle ABC$.

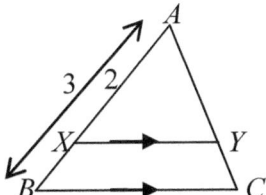

6. If the triangles below are similar, find the area of the smaller triangle.

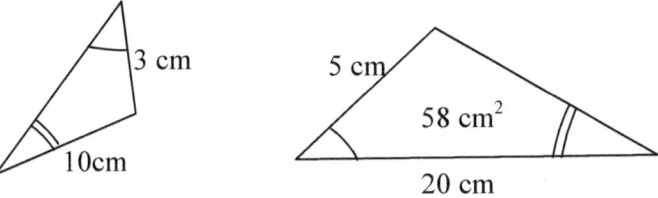

7. Two rectangles have areas 4 cm² and 9 cm² and their corresponding sides are x cm and 3 cm. Find the value of x.
8. In the following figure, AB is parallel to DC, $AB = 32$ cm, $BE = 12$ cm and $DE = 3$ cm. (a) Show that the triangles ABE and CDE are similar. Calculate: (b) The length CD. (c) The ratio of the areas of $\triangle ABE : \triangle CDE$.

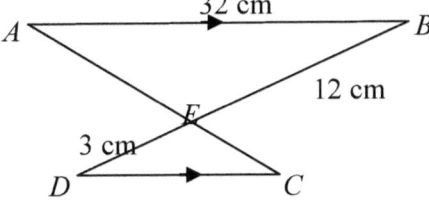

4.6 Ratio of Volumes of Similar Figures

Let the cylinder C_2 be an enlargement of C_1, with scale factor k.

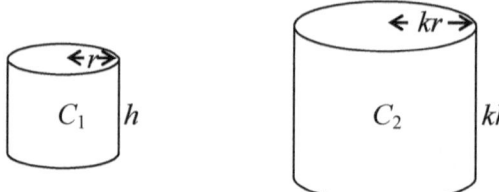

Volume of $C_1 = \pi r^2 h$ .... ①
Volume of $C_1 = \pi (kr)^2 (kh)$ .... ②
Dividing equation ② by equation ①

$$\frac{\text{Volume of } C_2}{\text{Volume of } C_1} = k^3$$

Therefore if the ratio of corresponding sides of two similar solid figures is k, then the ratio of their volumes is k^3.

 Example

A cube, whose volume is 20 cm³, is enlarged such that its volume now is V cm³. Given that, the scale factor of the enlargement is 4. Find the value of V.

Solution

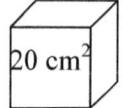   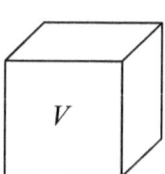

Let V_0 = volume of original cube

Then, $\dfrac{V}{V_0} = k^3 \Rightarrow V = k^3 V_0$

$k = 4$ and $V_0 = 20$ cm $\Rightarrow V = 4^3 \times 20 = 1280$ cm³

 Exercise 4:4

1. Two similar containers have heights of 3 cm and 5 cm respectively. If the capacity of the smaller container is 54 cm³, find the capacity of the larger container.
2. Two similar flasks are such that the ratio of the volume of the larger to the

smaller is 7:2. Calculate the volume of the larger flask if that of the smaller is 56 cm³. Given that the height of the smaller is 10 cm, find the height of the larger.

3. The following figure, shows two similar bowls with their volumes and the diameter of the smaller one given. Calculate the diameter d cm of the larger bowl.

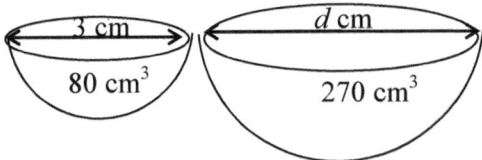

4. Two similar buckets have shapes in the form of a frustum. The volume of the smaller one is 24 cm³ and its slant height is 6 cm. Given that the larger one has a slant height of 9 cm, calculate the volume V of the larger one.

5. In the figure below, find the value of V, if the cones are similar.

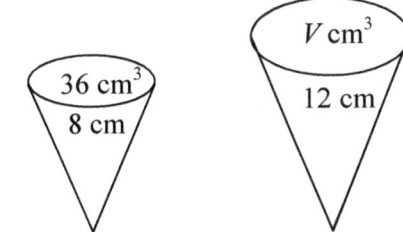

6. Given that the following cylinders are similar, find the value of y to the nearest tenth.

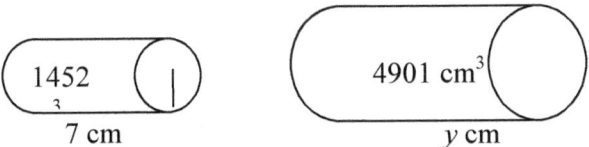

7. If the ratio of the diameters of two similar cups is 4:25, find the ratio of their volumes.

8. The surface area of a container of capacity 12.8 liters is 5000 cm². Find the surface area of a similar contain with a capacity of 5.5 liters.

Multiple Choice Exercise 4

1. The pair of triangles below which is definitely congruent is:

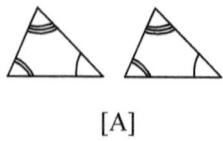
[A]

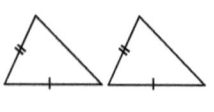
[B]

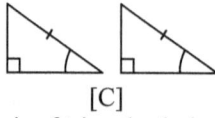
[C]

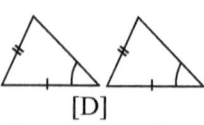
[D]

2. The pair of triangles below which is definitely congruent is:

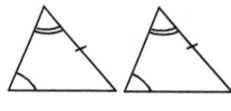
[A]

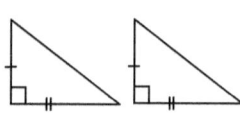
[B]

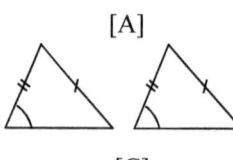
[C]

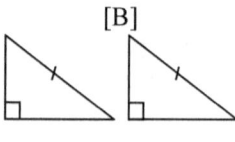
[D]

3. The triangles PQR and DEF are equiangular but not congruent. The triangles DEF and XYZ are congruent. It follows that:
   [A] $\triangle PQR$ and $\triangle DEF$ are equal in area
   [B] $\triangle PQR$ and $\triangle XYZ$ are congruent
   [C] $\triangle PQR$ and $\triangle XYZ$ are equal in area
   [D] $\triangle PQR$ and $\triangle XYZ$ are similar

4. The pair of triangles in below (not drawn to scale) which is similar is:

[A]

[B]

[C]

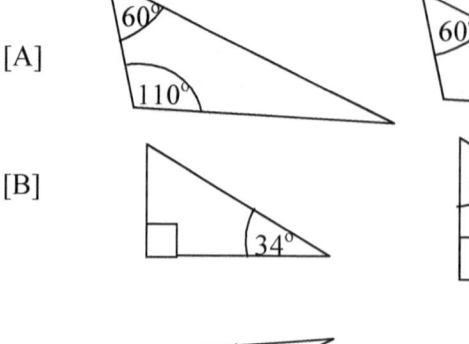
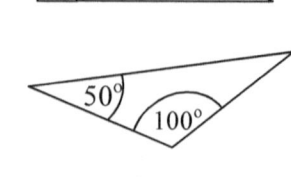

[D]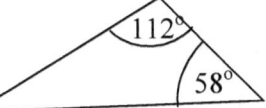

5. The pair of triangles below which is similar is:

   [A]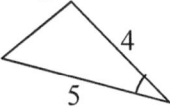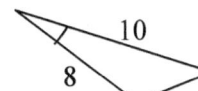

   [B]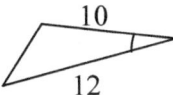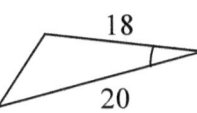

   [C]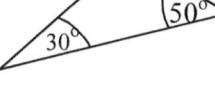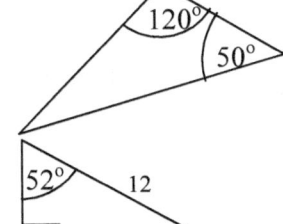

   [D]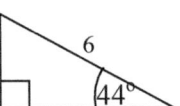

6. Given that the following triangles are similar. It follows that:

   [A] $\dfrac{AC}{XY}=\dfrac{YZ}{XZ}$ [B] $\dfrac{AC}{XY}=\dfrac{BC}{YZ}$ [C] $\dfrac{BC}{AB}=\dfrac{YZ}{XZ}$ [D] $\dfrac{BC}{AB}=\dfrac{XZ}{YZ}$

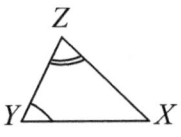

7. In the figure below, if $\dfrac{AB}{XY}=\dfrac{AC}{XZ}$ and $\angle B = \angle Y$.

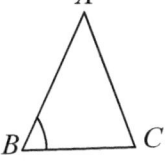 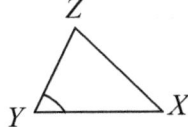

Then:

[A] $\dfrac{AB}{XY} = \dfrac{BC}{YZ}$ [B] $\Delta A = \Delta X$ [C] $\Delta C = \Delta Z$

[D] None of the above is necessarily true.

8. In the figure below, $\angle A = \angle X$ and $\angle B = \angle Y$. Hence, XY is equal to:

[A] $6\dfrac{7}{8}$ cm [B] $17\dfrac{3}{5}$ cm [C] $19\dfrac{1}{5}$ cm [D] $8\dfrac{1}{2}$ cm

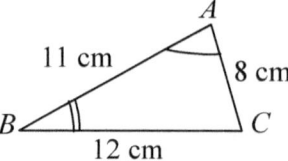
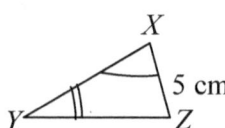

9. In the following figure, $PS = 8$ cm and $QS = 2$ cm. Hence $\dfrac{ST}{QR}$ is equal to:

[A] $\dfrac{1}{4}$ [B] $\dfrac{4}{1}$ [C] $\dfrac{4}{5}$ [D] $\dfrac{5}{4}$

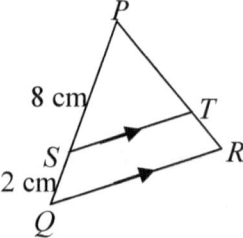

10. In the figure below, ST and QR are parallel. $|PS| = 6$ cm, $|SQ| = 8$ cm, $|PR| = 18\dfrac{2}{3}$ cm. $|PT|$ is equal to:

[A] 7 cm [B] 8 cm [C] $8\dfrac{2}{3}$ cm [D] 10 cm

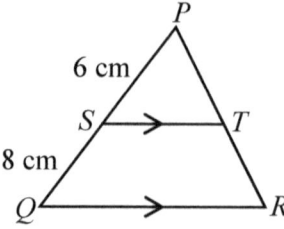

11. In the figure below, $|AB|=12$ cm, $|AE|=8$ cm, $|DC|=9$ cm and $AB\|DC$. The length $|EC|$ is:

Module 11, Topic 4: Congruency and Similarity

[A] 10 cm [B] 9 cm [C] 8 cm [D] 6 cm

12. In the following figure, ∠PMN=∠PRQ and ∠PNM=∠PQR. If |PM|=3 cm, |MQ|=7 cm and |PN|=5 cm, |NR|=

    [A] 1 cm [B] 3 cm [C] $3\frac{1}{2}$ cm [D] 5 cm

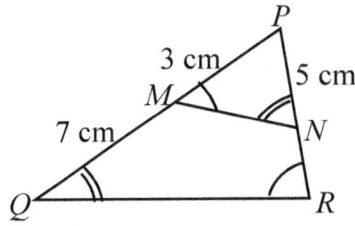

13. In the figure below, EF∥QR, PE = 2 cm, EQ = 4 cm and FR = 6 cm. x should be:
    [A] 2 cm [B] 3 cm [C] 4 cm [D] 6 cm

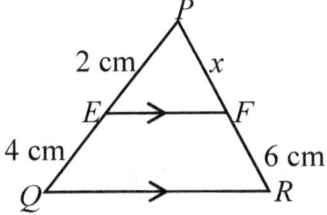

14. The value of x in the figure below is:
    [A] 6.8 cm [B] 6.6 cm [C] 6.5 cm [D] 5.6 cm

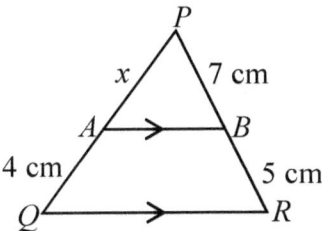

15. In the figure below, XY is parallel to BC and AB is parallel to YZ.

71

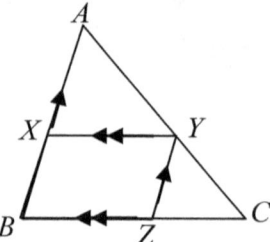

Therefore:

[A] $\angle ABC = \angle ZYC$ [B] $\dfrac{YZ}{ZC} = \dfrac{AC}{BC}$

[C] $\triangle ABC$ is similar to $\triangle ZYC$ [D] $\dfrac{ZC}{AC} = \dfrac{YZ}{AB}$

16. In the following figure AB is parallel to DC, $AB = 3$ cm and $DC = 5$ cm. Hence $\dfrac{XD}{XB}$ is equal to:

[A] $\dfrac{3}{5}$ [B] $\dfrac{5}{3}$ [C] $\dfrac{5}{8}$ [D] $\dfrac{8}{5}$

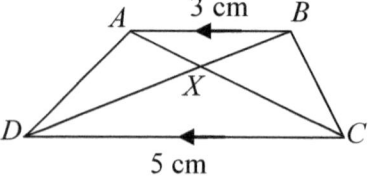

17. The triangles in below are:
[A] congruent [B] similar [C] identical [D] none of the above

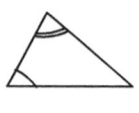   

18. The ratio of the areas of the two triangles shown below is:

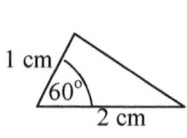   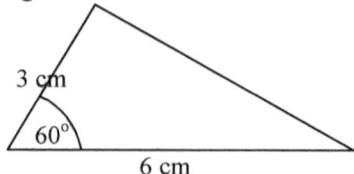

[A] 1:6   [B] 1:2   [C] 1:4   [D] 1:9

19. The following figure shows a right-angled triangle ACD.

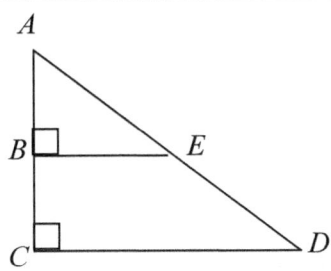

AB = 6 cm, AC = 8 cm, CD = 6 cm and BE is parallel to CD. The value of the length BE is:
    [A]   9   [B] 6   [C] 4.5   [D] 4

20. The following figure, shows two right-angled triangles OMN and OPQ. Given $\dfrac{OM}{OP} = \dfrac{1}{4}$, MN is parallel to PQ, $OM = 1$ cm, $ON = 2$ cm. The area of POQ in cm^2 is:
    [A] 20   [B] 16   [C] 12   [D] 8

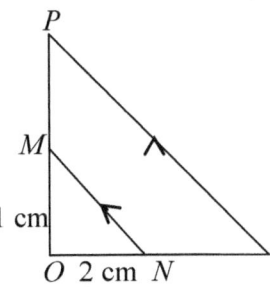

21. On a map drawn to a scale of 2 cm representing 1 km, the area represented by a square of side 4 cm is:
    [A] 2 km^2 [B] 4 km [C] 4 km^2 [D] 1 km^2

22. Two similar cylinders have heights of 3 cm and 6 cm respectively. The ratio of their volumes is:
    [A] 1:4   [B] 1:8   [C] 2:5   [D] 1:2

23. In following figure triangle ABC is similar to triangle AED and AB=16 cm, AE= 8 cm and AC=14 cm. The value of the length of the side marked x is:
    [A] 7 cm   [B] $\dfrac{80}{7}$ cm   [C] $\dfrac{70}{8}$ cm   [D] 6 cm

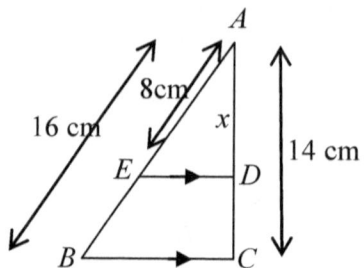

24. In the figure below, if the area of △XYZ is 10 cm², then the area of △ABC is:
   [A] 160 cm² [B] 40 cm² [C] 90 cm² [D] insufficient information.

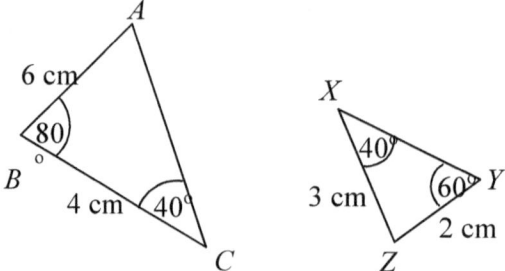

25. In the following figure, △ABC is similar to △DEF. Given that the area of △ABC is 20 cm², then the area of △DEF is:
   [A] 10 cm²   [B] 5 cm²   [C] 8 cm²   [D] None of the above

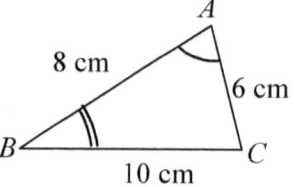
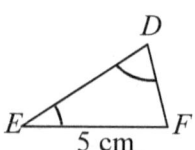

26. In the figure below, ∠A=∠X and ∠B=∠Y. △ABC has an area of 36 cm² and △XYZ has an area of 4 cm². If $AB = 4$ cm, then XY is equal to:
   [A] $\dfrac{3}{4}$ [B] $\dfrac{4}{3}$ [C] $\dfrac{4}{4}$ [D] $\dfrac{9}{4}$

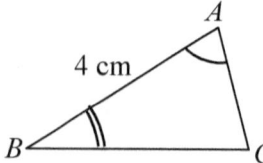
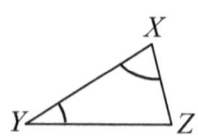

27. Two buckets A and B, identical in shape, are such that the dimensions of A are three times as large as the corresponding dimensions of B. The ratio of the volumes of A:B is:
   [A] 1:9   [B] 27:1   [C] 9:1   [D] 1:27
28. The pairs of triangles PQR, XYZ are congruent is:

[A] $XY = PQ, XZ = QR, \angle X = \angle Q$
[B] $XZ = QR, YZ = PR, \angle Y = \angle P$
[C] $\angle Y = \angle P, \angle Z = \angle Q, XZ = PQ$
[D] $\angle Z = \angle P, \angle Y = \angle Q, XY = PR$

29. Similar triangles differ from congruent triangles in that:
    [A] The areas of congruent triangles are not necessarily equal but the areas of similar triangles are necessarily equal.
    [B] The areas of congruent triangles are necessarily equal but the areas of similar triangles are not necessarily equal.
    [C] Similar triangles are necessarily congruent but congruent triangles are not necessarily similar.
    [D] The sides of similar triangles are necessarily equal but the sides of congruent triangles are not necessarily equal.

30. The set, which consist entirely of elements, which are similar figures is:
    [A] Triangles [B] Quadrilaterals [C] Circles [D] Hexagons

31. The false statement (s) is/are:
    [A] All Similar objects have the same shape but not necessarily the same size.
    [B] All similar objects are congruent.
    [C] All congruent objects are similar.
    [D] All similar objects are congruent and have the same shape but not necessarily the same size.

Topic 5

VECTORS

Objectives
At the end of this topic, the learner should be able to:

1. Distinguish between vector and scalar quantities.
2. Notate vectors and represent vectors in various ways.
3. Find the magnitude or norm of a vector.
4. Distinguish between fixed, free vectors and position vectors.
5. Carry out vector addition, subtraction and multiplication by a scalar quantity.
6. State and use the conditions for vectors to be equal.

Module 11, Topic 5: Vectors

5.1 Vector and Scalar Quantities

> **? Brainstorming Exercise**
>
> The following figure, is a map of a village of four quarters, namely; north quarter N, west quarter W, south quarter S and east quarter E. A straight road links N and S and another links W and E in such a way that they meet at the junction J. Each quarter is 4 km away from J.
>
>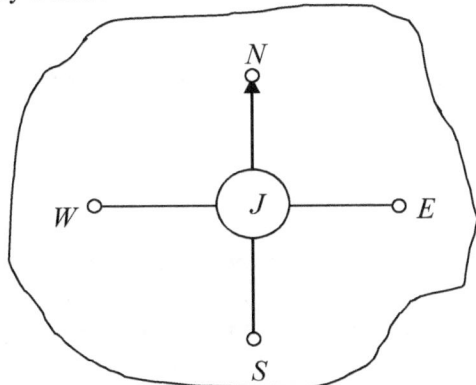
>
> Three boys Nfor, Kanjo and Shey while standing at J, describe the journeys they will undertake. Nfor says "I will travel 4 km from here". Kanjo says "I will travel 4 km along the road linking W and E". Shey says "I will travel 4 km towards E".
>
> Answer the following question giving reasons for your answers.
>
> 1. What distance will each of them cover?
> 2. To which quarter do you think each of them is going to?

From the statements they make it is clear that each of them will cover a **distance** of 4 km. 4 km is called the **magnitude, size, length, modulus or norm** of the journey.

One cannot be able to say exactly to which quarter Nfor will be going. From Kanjo's statement, he will be going to either W or E, though he has not stated precisely whether he will go to W or he will go to E. Kanjo has mentioned his **distance** and **direction**. If Kanjo travels from J to W, his direction is the same as if he travels from J to E.

It is certain that Shey will travel to E, since he is traveling towards E and is going to cover a distance of 4 km. Shey's description towards E" gives the **sense** of his journey.

Displacement is the distance covered in a specific direction.

In the statement made by Shey above, there are three pieces of information.

1. The **length, magnitude, size, norm** or **modulus** is 4 km.
2. The **direction** is along the road joining W and E.
3. The **sense** is towards E.

A **vector quantity** or simply a **vector** is a quantity carry, which has magnitude, direction and sense. Examples of vector quantities are displacement, velocity, momentum, force, acceleration etc.

A **scalar quantity** or **scalar** on the other hand has no direction and hence no sense. Examples of scalar quantities are mass, temperature, distance, speed, area, length, volume, time, age marks, price etc.

5.2 Vectors as Directed Line Segments

If L is a straight line and X and Y are any two points on the line, then the part of the line between and including X and Y is a **line segment**. Therefore, *a line segment is simply part of a line*. A **directed line segment** is a line segment with an arrow at one end. Diagrammatically, we can represent vectors using directed line segments. We call the end carrying the arrow the **head** and the other end the **tail** of the vector.

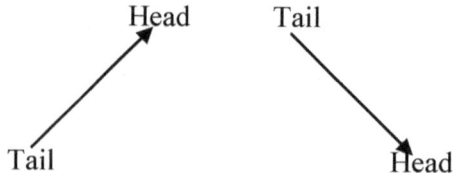

The length of a directed line segment, gives the magnitude of the vector, the orientation of the line gives the direction while the arrow gives the sense of the vector.

5.3 Vector Notation

We denote a vector represented by the directed line segment AB as in the figure below, $\underline{a}$, $\vec{a}$ or $\overrightarrow{AB}$. The arrow emphasizes that the sense is from A to B. Textbooks use bold print. Thus we denote the vector in the following figure by **AB** or **a**. In using bold print, arrows or bars are not necessary.

$A \longrightarrow B$

We denote the magnitude of the vector **AB** or **a**, above by $|\underline{a}|, |\vec{a}|, |\overrightarrow{AB}|$ or AB and define it as the length of the line segment from A to B.

Module 11, Topic 5: Vectors

5.4 Column Vectors

In column vector form, we write a vector **AB** that is a units in the Ox direction and b units in the Oy direction as $\begin{pmatrix} a \\ b \end{pmatrix}$.

We call a the "x **component** of **AB**" and call b the y **component** of **AB**".

 Example

Write the vectors on the Cartesian plane below as column vectors.

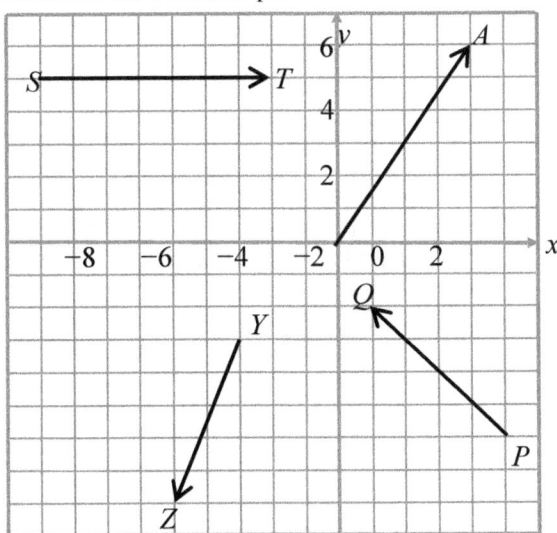

Solution

$\overrightarrow{OA} = \begin{pmatrix} 4 \\ 6 \end{pmatrix}, \overrightarrow{ST} = \begin{pmatrix} 7 \\ 0 \end{pmatrix}, \overrightarrow{YZ} = \begin{pmatrix} -2 \\ -5 \end{pmatrix}, \overrightarrow{PQ} = \begin{pmatrix} -4 \\ 4 \end{pmatrix}$

 Exercise 5:1

1. Represent the following on a Cartesian plane.

   $\mathbf{OA} = \begin{pmatrix} 5 \\ 2 \end{pmatrix}$, $\mathbf{OB} = \begin{pmatrix} -5 \\ 3 \end{pmatrix}$, $\mathbf{OC} = \begin{pmatrix} -4 \\ 1 \end{pmatrix}$, $\mathbf{OD} = \begin{pmatrix} 6 \\ -2 \end{pmatrix}$

2. Write down the vectors in the diagram below as column vectors.

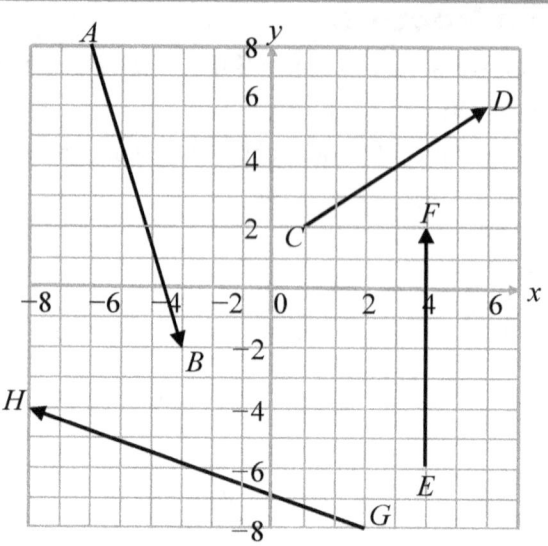

3. Given that A, B, C, D and E are the points $(2,5), (-1,4), (0,3), (-5,-6)$ and $(7,-2)$ respectively. Represent the following vectors on a Cartesian plane.

$$AV = \begin{pmatrix} 2 \\ 4 \end{pmatrix}, BW = \begin{pmatrix} -3 \\ 5 \end{pmatrix}, CX = \begin{pmatrix} -2 \\ -3 \end{pmatrix}, DY = \begin{pmatrix} 0 \\ 6 \end{pmatrix}, EZ = \begin{pmatrix} -5 \\ 0 \end{pmatrix}$$

4. Given that A, B, C, D, E are the points $(2, 5), (-1, 4), (0, 3), (-5, -6)$, and $(7, -2)$ respectively. Represent the vectors **AB**, **CE** and **BD** on a Cartesian plane.

5.5 The Magnitude or Modulus of a Vector

We can write the column vector $\begin{pmatrix} x \\ y \end{pmatrix}$ in **component form** as $x\mathbf{i} + y\mathbf{j}$. **i** and **j** are perpendicular unit base vectors.

We can use the Pythagoras theorem to find the magnitude of **AB** as below.

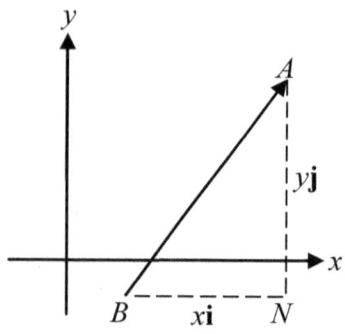

Thus, $|\mathbf{AB}| = \sqrt{x^2 + y^2}$.

Note!
The magnitude of a vector is always positive.

If the end points of the vector **AB** are $A(x_1, y_1)$ and $B(x_2, y_2)$ the modulus will be the distance between the points A and B.

Thus, $AB = \sqrt{(x_2 - x_1)^2 + (y_2 - y_1)^2}$

Example

1. Find the modulus of the following vectors:

   (i) $PQ = \begin{pmatrix} -4 \\ 3 \end{pmatrix}$ (ii) $\mathbf{XY} = 5\mathbf{i} - 12\mathbf{j}$

   Solution

   (i) $PQ = \sqrt{(-4)^2 + 3^2} = \sqrt{25} = 5$ units

   (ii) $\mathbf{XY} = \sqrt{5^2 + (-12)^2} = \sqrt{169} = 13$ units

2. Calculate the magnitude of the vector **YZ** where Y and Z are the points $(-2, 2)$ and $(4, -6)$ respectively.

   Solution
   $YZ = \sqrt{(-6-2)^2 + (4-(-2))^2} = \sqrt{100} = 10$ units

Exercise 5:2

1. Find the magnitudes of the vectors:

   (i) $\begin{pmatrix} 5 \\ 12 \end{pmatrix}$ (ii) $2\mathbf{i} - 3\mathbf{j}$ (iii) $\begin{pmatrix} -7 \\ 3 \end{pmatrix}$ (iv) $2\mathbf{i} - 3\mathbf{j}$ (v) $\begin{pmatrix} -8 \\ -6 \end{pmatrix}$ (vi) $-4\mathbf{i} + 3\mathbf{j}$

2. Calculate the modulus of the vectors joining the given points.

   (i) $(3, 5)$ and $(-2, 1)$ (ii) $(-4, 7)$ and $(0, 3)$ (iii) $(-2, -1)$ and $(-5, -3)$

3. Which of the following are unit vectors?

   (i) $\mathbf{i}+\mathbf{j}$ (ii) $\mathbf{i}$ (iii) $\frac{1}{2}\mathbf{i}+\frac{1}{2}\mathbf{j}$ (iv) $\frac{3}{5}\mathbf{i}+\frac{4}{5}\mathbf{j}$

   (v) $\mathbf{j}$ (vi) $\frac{1}{\sqrt{2}}\mathbf{i}-\frac{1}{\sqrt{2}}\mathbf{j}$ (vii) $-\frac{\sqrt{3}}{2}\mathbf{i}+\frac{1}{2}\mathbf{j}$

4. Given that $\mathbf{u}=-3\mathbf{i}+4\mathbf{j}$. Find the magnitude of **u**.
5. Find $|\mathbf{V}|$ if $\mathbf{V}=3\mathbf{i}-2\mathbf{j}$.

VECTOR ALGEBRA

5.6 Equality of Vectors

Two vectors **a** and **b** are equal if and only if they satisfy the following conditions.
(i) They have the same magnitude
(ii) They are in the same direction (parallel or collinear)
(iii) They have the same sense.

Example

In the figure below, which of the vectors are equal?

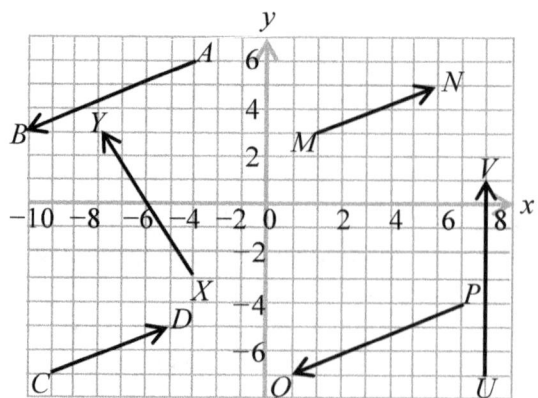

Solution

$$AB = PQ = \begin{pmatrix} -7 \\ -3 \end{pmatrix} \qquad CD = MN = \begin{pmatrix} 5 \\ 2 \end{pmatrix}$$

5.7 Fixed and Free Vectors

From the figure above, we see that though the vectors **AB** and **PQ** are equal, we can represent them at different positions on a Cartesian plane. This is equally true of **CD** and **MN**. We call a vector which we can situate anywhere in a plane a **free vector**. Free vectors have no particular point of action and only their magnitude, direction and sense is important. The point of action of some other vectors is very important. These types of vectors therefore have a fixed position in a plane or space and we call them **fixed vectors**. A practical example of a fixed vector is the gravitation force acting on an object. The point of action of this force is the center of mass of the object and the direction and sense is towards the center of the earth.

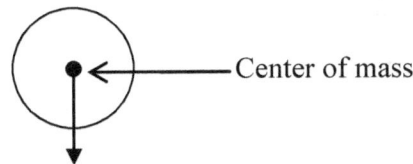

Gravitational force

Another example is a vector **EF** with given end points such as $E(2, 3)$ and $F(-7, 4)$.

Position Vectors

If $P(x, y)$ is any point in the x-y plane, then we denote the position vector of the point P by **r** given by

$$\mathbf{r} = \begin{pmatrix} x \\ y \end{pmatrix} \text{ or } \mathbf{r} = x\mathbf{i} + y\mathbf{j}$$

A position vector is another example of a fixed vector discussed earlier and is never a free vector. This is because the tail of a position vector is always at the origin. This means that tail must be at the origin.

 Example

Draw the position vectors of the points $A(3,1)$ and $B(-5,4)$ on a Cartesian plane.

Solution

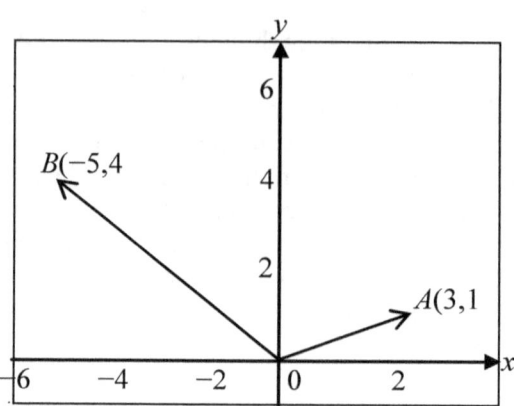

Write the position vector of the point $H(-7, 5)$ in
(a) Column vector form (b) Component form.

Solution

(a) In column vector form, $\mathbf{r} = \begin{pmatrix} -7 \\ 5 \end{pmatrix}$ (b) In component form, $\mathbf{r} = -7\mathbf{i} + 5\mathbf{j}$.

 Exercise 5:4

1. Write the position vectors of the points labeled A to F in:
   (a) Component form (b) Column vector form.

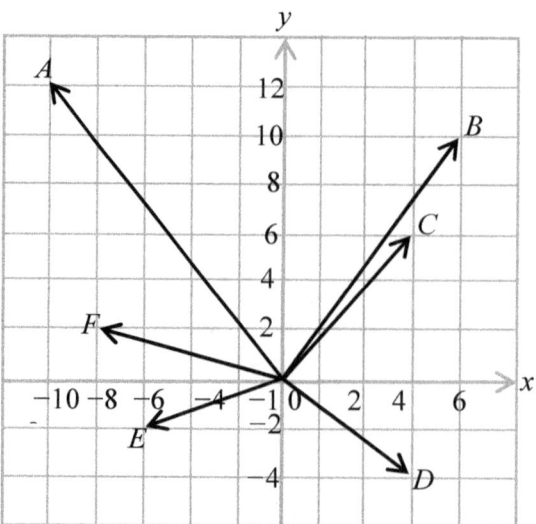

2. (i) Write the position vector of the points $A(-1,5)$, $B(4,-7)$, $C(-9, 3)$ and $D(-3, 6)$
   (a) Component form (b) Column vector form.
   (ii) Represent these position vectors on the Cartesian plane.
3. In the following figure, which of the vectors are equal?

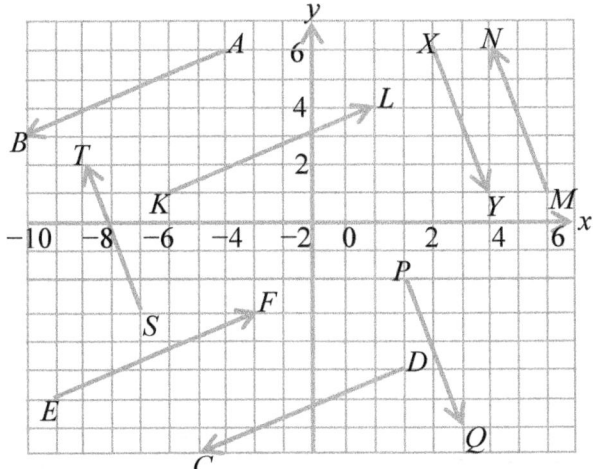

4. Which of the following vectors are equal?
   (a) $-5\mathbf{i} - 2\mathbf{j}$ (b) $2\mathbf{i} + 5\mathbf{j}$ (c) $-2\mathbf{i} - 5\mathbf{j}$ (d) $5\mathbf{i} + 2\mathbf{j}$
   (e) $2(-5\mathbf{i} - 2\mathbf{j})$ (f) $4(2\mathbf{i} + 5\mathbf{j})$ (g) $-1(-2\mathbf{i} - 5\mathbf{j})$ (h) $3(5\mathbf{i} + 2\mathbf{j})$
   (i) $\begin{pmatrix} -2 \\ -5 \end{pmatrix}$ (j) $\begin{pmatrix} 2 \\ 5 \end{pmatrix}$ (k) $\begin{pmatrix} 5 \\ 2 \end{pmatrix}$ (l) $\begin{pmatrix} -5 \\ -2 \end{pmatrix}$ (m) $-1\begin{pmatrix} -2 \\ -5 \end{pmatrix}$
   (n) $3\begin{pmatrix} 2 \\ 5 \end{pmatrix}$ (o) $4\begin{pmatrix} 5 \\ 2 \end{pmatrix}$ (p) $2\begin{pmatrix} -5 \\ -2 \end{pmatrix}$
5. In the figure below, which of the vectors are position vectors relative to the origin?

Competency Based Mathematics for Secondary Schools. Book 3

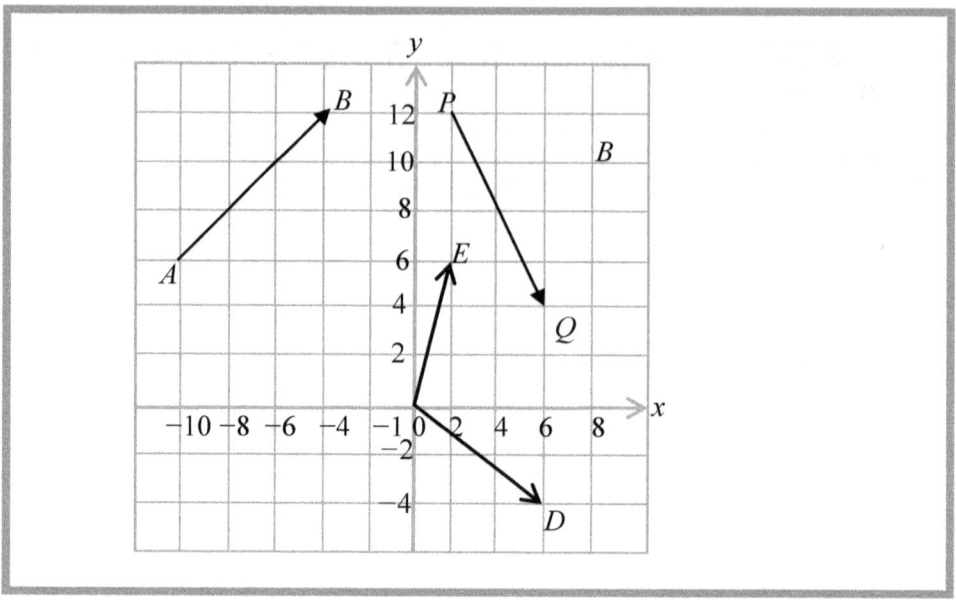

5.8 Addition of Vectors

1. To add vectors written in column vector form we add corresponding entries just as in matrix addition.
2. To add vectors written in component form, we add like components.

 Example

6. Find $\mathbf{x} + \mathbf{y}$ if $\mathbf{x} = \begin{pmatrix} 3 \\ 4 \end{pmatrix}$ and $\mathbf{y} = \begin{pmatrix} 1 \\ -2 \end{pmatrix}$.

   Solution

   $\mathbf{x} + \mathbf{y} = \begin{pmatrix} 3 \\ 4 \end{pmatrix} + \begin{pmatrix} 1 \\ -2 \end{pmatrix} = \begin{pmatrix} 4 \\ 2 \end{pmatrix}$

7. If $\mathbf{a} = 3\mathbf{i} + 4\mathbf{j}$ and $\mathbf{b} = \mathbf{i} + 2\mathbf{j}$, find $\mathbf{a} + \mathbf{b}$.

   Solution
   $\mathbf{a} + \mathbf{b} = (3\mathbf{i} + 4\mathbf{j}) + (\mathbf{i} + 2\mathbf{j}) = 4\mathbf{i} + 6\mathbf{j}$

Module 11, Topic 5: Vectors

Adding Vectors Diagrammatically

Diagrammatically we add vectors by drawing to scale the vectors in such a way that the head of one vector is at the tail of the other. The sum is then the vector whose tail is at the tail of the first vector and whose head is at the head of the second vector.

Example

Find the sum of the vectors **a** and **b** shown diagrammatically below.

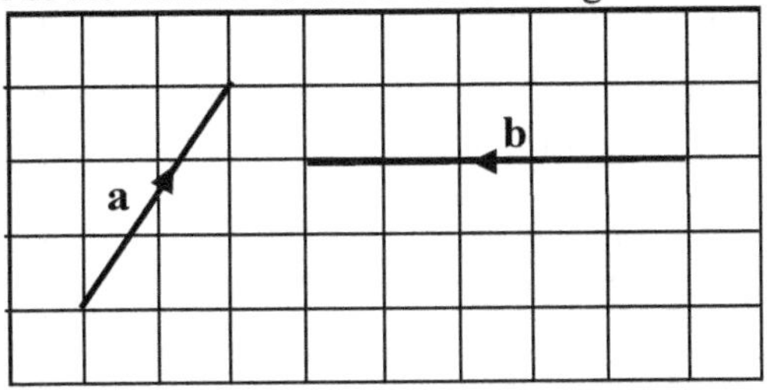

Solution

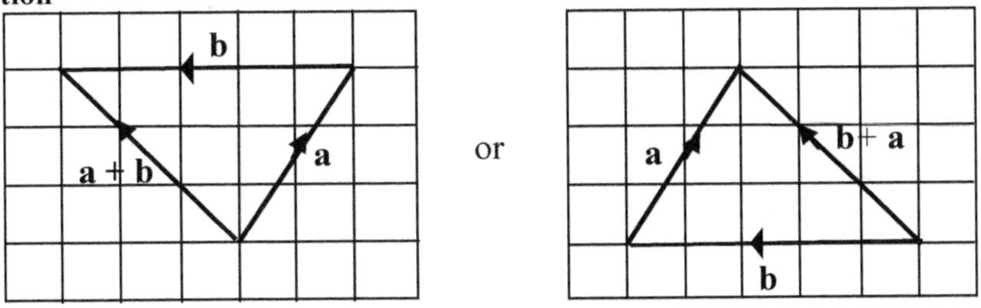

We then obtain the sum by completing and measuring the completed side of the triangle called the **triangle of vectors**.

By measuring **a** + **b** and **b** + **a** from the two triangles above we can see that **a** + **b** = **b** + **a**. This means that vector **addition is commutative**.

The polygon of vectors

We can extend the idea of the triangle of vectors to any other polygon, depending on the number of vectors we have to add.

Example

Use the following figure to show that **AE** = **a** + **b** + **c** + **d**.

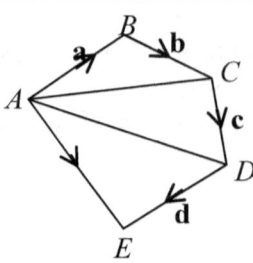

Solution

$$AC = AB + BC \Rightarrow AC = a + b$$
$$AD = AC + CD = a + b + c$$
$$AE = AD + DE = a + b + c + d$$

Hence, $AB + BC + CD + DE = AE$.

The Additive Inverse of a Vector

Consider the vector **AB** in figure (i). We can reverse the sense of **AB** to obtain **BA** as in figure (ii).

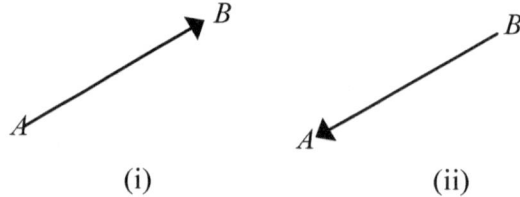

BA is called the **additive inverse** (or the **opposite**) of **AB**.
Therefore, $AB = -BA$

The additive inverse of **a** is **−a**

The additive inverse of $\begin{pmatrix} x \\ y \end{pmatrix}$ is $\begin{pmatrix} -x \\ -y \end{pmatrix}$.

The additive inverse of $a\mathbf{i} + b\mathbf{j}$ is $-a\mathbf{i} - b\mathbf{j}$.

5.9 Subtraction of Vectors

Subtracting Column vectors
To subtract vectors written in column vector form, we subtract corresponding entries.

Module 11, Topic 5: Vectors

 Example

Find $\mathbf{a} - \mathbf{b}$ if $\mathbf{a} = \begin{pmatrix} 4 \\ 1 \end{pmatrix}$ and $\mathbf{b} = \begin{pmatrix} 7 \\ 3 \end{pmatrix}$.

Solution

$\mathbf{a} - \mathbf{b} = \begin{pmatrix} 4 \\ 1 \end{pmatrix} - \begin{pmatrix} 7 \\ 3 \end{pmatrix} = \begin{pmatrix} -3 \\ -2 \end{pmatrix}$

Subtracting component form Vectors

To subtract vectors written in component form, we subtract like components.

 Example

If $\mathbf{a} = 7\mathbf{i} + 3\mathbf{j}$ and $\mathbf{b} = 4\mathbf{i} - \mathbf{j}$. Find $\mathbf{a} - \mathbf{b}$.

Solution
$\mathbf{a} - \mathbf{b} = (7\mathbf{i} + 3\mathbf{j}) - (4\mathbf{i} - \mathbf{j}) = 3\mathbf{i} + 4\mathbf{j}$

Subtracting Vectors Diagrammatically

Subtraction of vectors is the same as the addition of inverses. This means that $\mathbf{a} - \mathbf{b} = \mathbf{a} + (-\mathbf{b})$.

We can use this idea to subtract vectors diagrammatically.

 Example

The following figure shows the representation of the vectors **a** and **b**. Draw a diagram to show the vector $\mathbf{a} - \mathbf{b}$.

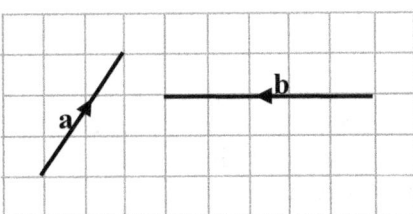

Solution
The figure below shows the vector −b.

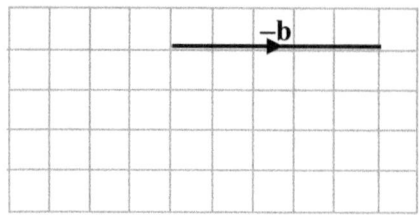

The following shows how to add **a** and −**b** diagrammatically.

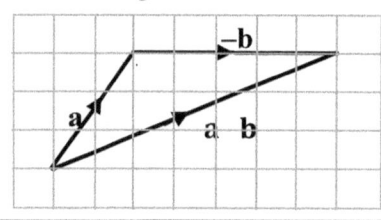   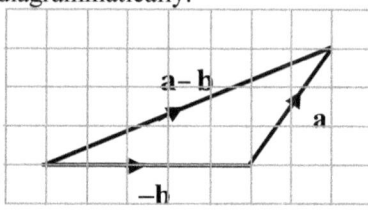

Or

The Zero or Null Vector

Consider two vectors **a** and **b** such that **a** = **b**
Then **a** − **b** = **a** + (− **b**) = **a** + (− **a**)
But **a** − **b** = **a** − **a** = **0**
∴ **a** − **a** = **0**

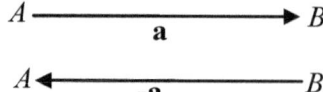

a + (− **a**) therefore, means a journey from A to B and back to A, or simply a zero displacement. We call such a vector a **zero vector** or **null vector** denoted by $\mathbf{0} = \begin{pmatrix} 0 \\ 0 \end{pmatrix}$.

The Parallelogram Law of Vectors

If two vectors **OA** = **a** and **OC** = **c** have a common point of action O (i.e. if their tails are at the same point O) but are acting in different directions as shown in the diagram below, then we can obtain their sum and difference by completing the parallelogram $OABC$ and measuring its diagonals.

Module 11, Topic 5: Vectors

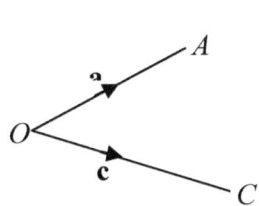 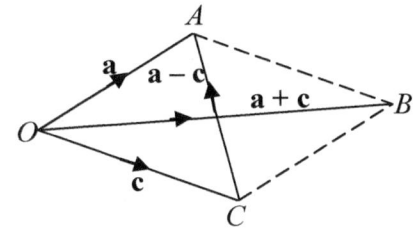

The diagonal *AC* joining the heads represents the difference while the diagonal *OB*, represents the sum of the vectors.
Thus **a + c = OB** and **a − c = CA**

The sense of the difference **CA = a − c** is obtained by noting that **− c** means the additive inverse **+ (−c)** of **c**.

Exercise 5:5

1. The figure below shows two vectors **a** and **b**. Illustrate on different Cartesian planes how you would diagrammatically find:
   (i) **a + b**   (ii) **a − b**   (iii) **b − a**

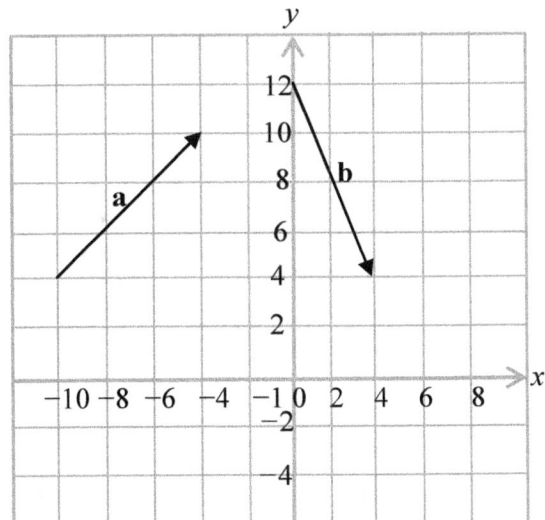

2. If $\mathbf{u} = 3\mathbf{i} - \mathbf{j}$ and $\mathbf{v} = -5 + 3\mathbf{j}$, find
   (i) **u + v**   (ii) **v + u**   (iii) **u − v**   (iv) **v − u**
3. Give the additive inverse of each of the following vectors:
   (a) $-\mathbf{i} - 2\mathbf{j}$ (b) $2\mathbf{i} + 5\mathbf{j}$ (c) $2\mathbf{i} - 7\mathbf{j}$ (d) $-\mathbf{i} + 2\mathbf{j}$

(e) $\begin{pmatrix} -2 \\ -4 \end{pmatrix}$ (f) $\begin{pmatrix} 7 \\ 5 \end{pmatrix}$ (g) $\begin{pmatrix} -3 \\ 2 \end{pmatrix}$ (h) $\begin{pmatrix} 8 \\ -3 \end{pmatrix}$

4. Given the vectors $\mathbf{a} = \begin{pmatrix} 2 \\ 5 \end{pmatrix}$ and $\mathbf{b} = \begin{pmatrix} -3 \\ 2 \end{pmatrix}$, evaluate:

   (i) $\mathbf{a} + \mathbf{b}$ (ii) $\mathbf{b} + \mathbf{a}$ (iii) $\mathbf{a} - \mathbf{b}$ (iv) $\mathbf{b} - \mathbf{a}$
   (v) What conclusions do you draw from your results in (i), (ii), (iii) and (iv)?

5.10 Multiplication of Vectors by Scalars

We can consider multiplication of a vector by a scalar as repeated addition of the same vector. Thus if $\mathbf{d}$ is any vector then $5\mathbf{d} = \mathbf{d} + \mathbf{d} + \mathbf{d} + \mathbf{d} + \mathbf{d}$.

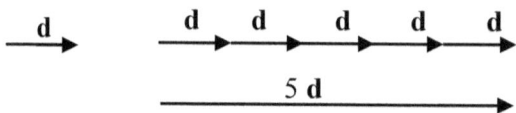

Generally for any scalar n, $n\mathbf{d} = \mathbf{d} + \mathbf{d} + \mathbf{d} + \ldots +$ up to n times.

This generalization is true even if n is a fraction or n is negative. If n is negative, the sense of the vector will be opposite to the sense of the original vector. Therefore, if $\mathbf{a}$ and $\mathbf{b}$ are parallel vectors then we can express $\mathbf{a}$ as a scalar multiple of $\mathbf{b}$. i.e. $\mathbf{a} = n\mathbf{b}, n \in \mathbb{R}$.

 Exercise 5:6

1. Given that $\mathbf{u} = \begin{pmatrix} 2 \\ 3 \end{pmatrix}$ and $\mathbf{v} = \begin{pmatrix} 0 \\ 1 \end{pmatrix}$, find numbers a and b such that $a\mathbf{u} + b\mathbf{v} = \begin{pmatrix} 4 \\ 5 \end{pmatrix}$.

2. Given that $\mathbf{a} = \begin{pmatrix} 4 \\ -3 \end{pmatrix}$, $\mathbf{b} = \begin{pmatrix} 5 \\ 7 \end{pmatrix}$ and $\mathbf{c} = \begin{pmatrix} -6 \\ 4 \end{pmatrix}$, evaluate the following.

   (i) $\mathbf{a} + \mathbf{b}$ (ii) $\mathbf{a} - \mathbf{b}$ (iii) $\mathbf{b} - \mathbf{a}$ (iv) $\mathbf{a} + (\mathbf{b} + \mathbf{c})$
   (v) $(\mathbf{a} + \mathbf{b}) + \mathbf{c}$ (vi) $3\mathbf{a}$ (vii) $\mathbf{b} - \dfrac{1}{2}\mathbf{c}$ (viii) $\mathbf{b} - 2\mathbf{a} + 3\mathbf{c}$
   (ix) $\dfrac{1}{3}(3\mathbf{a} + 2\mathbf{a} + \mathbf{c})$ (x) $2\mathbf{a} + \mathbf{b} - 3\mathbf{c}$

Module 11, Topic 5: Vectors

3. Given the vectors $\mathbf{x} = \begin{pmatrix} 2 \\ 1 \end{pmatrix}, \mathbf{y} = \begin{pmatrix} 1 \\ -4 \end{pmatrix}, \mathbf{z} = \begin{pmatrix} 1 \\ -1 \end{pmatrix}$, find the vector $\mathbf{x} + \mathbf{y}$ and give a reason why this vector is in the same direction as $\mathbf{z}$.

4. The position vectors of A, B, C and D are respectively $\mathbf{a}$, $\mathbf{b}$, $\mathbf{c}$ and $\mathbf{d}$, where $2\mathbf{c} = \mathbf{a}$ and $4\mathbf{d} = \mathbf{a} + \mathbf{b}$.
   (a) Show that $\mathbf{AB}$ is parallel to $\mathbf{CD}$. (b) Find the ratio $AB : CD$.

5. Given that $\mathbf{a} = 3\mathbf{i} - 4\mathbf{j}$ and $\mathbf{b} = 4\mathbf{i} + 3\mathbf{j}$, find the magnitude of $\mathbf{a} + \mathbf{b}$, leaving your answer in surd form.

6. Given that $\mathbf{a} = \begin{pmatrix} 3 \\ 4 \end{pmatrix}, \mathbf{b} = \begin{pmatrix} 1 \\ 4 \end{pmatrix}$ and $\mathbf{c} = \begin{pmatrix} 5 \\ 12 \end{pmatrix}$ and that $u\mathbf{a} + v\mathbf{b} = \mathbf{c}$. Write simultaneous equations in u and v and solve them.

Multiple Choice Exercise 5

1. The modulus of $6\mathbf{i} + 8\mathbf{j}$ is:
   [A] $2\sqrt{7}$ [B] 8   [C] 10   [D] 6

2. If $\mathbf{a} = 3\mathbf{i} - \mathbf{j}$ and $\mathbf{b} = \mathbf{i} + 2\mathbf{j}$, then $\mathbf{a} \cdot \mathbf{b}$ is equal to:
   [A] 1   [B] 5   [C] –5   [D] –1

3. Given that A, B and C are collinear and $\mathbf{OA} = \mathbf{i} + \mathbf{j}$, $\mathbf{OB} = 2\mathbf{i} - \mathbf{j}$ and $\mathbf{OC} = 3\mathbf{i} + a\mathbf{j}$. The value of a is:
   [A] 1   [B] –1   [C] –3   [D] –2

4. The vector $\begin{pmatrix} -4 \\ 1 \end{pmatrix}$ in $\mathbf{i}$ and $\mathbf{j}$ component form is:
   [A] $4\mathbf{i} - \mathbf{j}$ [B] $\mathbf{i} - 4\mathbf{j}$ [C] $-4\mathbf{i} + \mathbf{j}$ [D] $-\mathbf{i} - 4\mathbf{j}$

5. The vector $2\mathbf{i} - 3\mathbf{j}$ in column vector form is:
   [A] $\begin{pmatrix} -3 \\ -2 \end{pmatrix}$ [B] $\begin{pmatrix} 3 \\ -2 \end{pmatrix}$ [C] $\begin{pmatrix} -2 \\ 3 \end{pmatrix}$ [D] $\begin{pmatrix} 2 \\ -3 \end{pmatrix}$
   (a)

6. The additive inverse of $3\mathbf{i} - 5\mathbf{j}$ is:
   [A] $3\mathbf{i} + 5\mathbf{j}$ [B] $-3\mathbf{i} + 5\mathbf{j}$ [C] $-3\mathbf{i} - 5\mathbf{j}$ [D] $3\mathbf{i} - 5\mathbf{j}$

7. The additive inverse of $\begin{pmatrix} -3 \\ 2 \end{pmatrix}$ is:
   [A] $\begin{pmatrix} -3 \\ -2 \end{pmatrix}$ [B] $\begin{pmatrix} 3 \\ -2 \end{pmatrix}$ [C] $\begin{pmatrix} -2 \\ 3 \end{pmatrix}$ [D] $\begin{pmatrix} 2 \\ -3 \end{pmatrix}$

8. As a column vector, the vector represented by the directed line segment in

the figure below is:

[A] $\begin{pmatrix} -3 \\ 8 \end{pmatrix}$ [B] $\begin{pmatrix} 3 \\ -8 \end{pmatrix}$ [C] $\begin{pmatrix} -8 \\ 3 \end{pmatrix}$ [D] $\begin{pmatrix} 8 \\ -3 \end{pmatrix}$

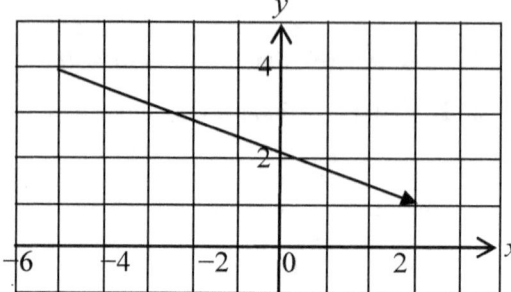

9. On the Cartesian plane, the free vector $-5\mathbf{i} + 3\mathbf{j}$ is represented as:

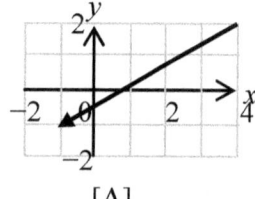 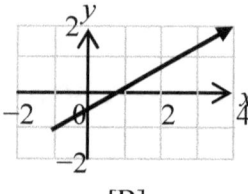

[A] [B]

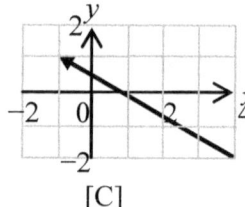 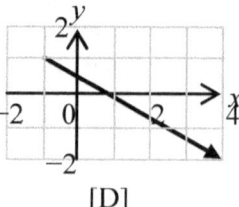

[C] [D]

10. Given that the position vectors of A, B, C and D are respectively **a**, **b**, **c** and **d** where, $2\mathbf{c} = \mathbf{a}$ and $4\mathbf{d} = \mathbf{a} + \mathbf{b}$. We can write **CD** in terms of **AB** as:
    [A] $4\mathbf{CD} = \mathbf{AB}$ [B] $4\mathbf{CD} = -\mathbf{AB}$
    [C] $\mathbf{CD} = 4\mathbf{AB}$ [D] $\mathbf{CD} = -4\mathbf{AB}$

11. In the following figure, the pair of vectors, which are parallel, are:
    [A] **AB** and **NM** [B] **XY** and **MN** [C] **XY** and **PQ** [D] **BA** and **MN**

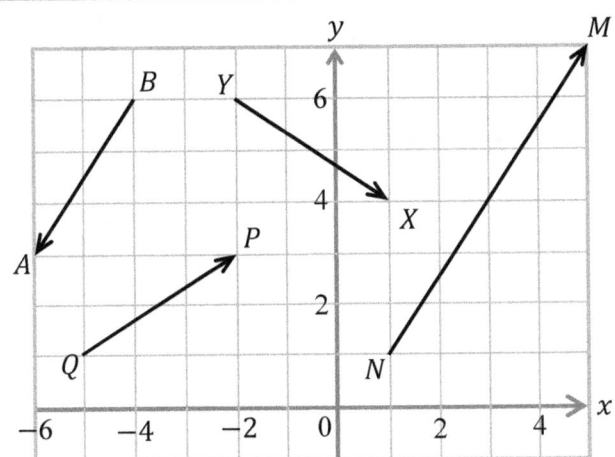

12. Given that, **a** = −5i+12j, **b** = 12i−5j, **c** =5i −12j and **d** = $\begin{pmatrix} -5 \\ 12 \end{pmatrix}$. The vectors, which are equal, are:
    [A] **a** and **b**   [B] **a** and **c**   [C] **a** and **d**   [D] **c** and **d**

13. Given that, **a** = $\begin{pmatrix} -9 \\ 12 \end{pmatrix}$, **b** = $\begin{pmatrix} -12 \\ 9 \end{pmatrix}$, **c** = −9i+12j and **d** = 12i−9j. The vectors, which are equal, are:
    [A] **a** and **b**   [B] **a** and **c**   [C] **a** and **d**   [D] **b** and **d**

14. The directed line segments [AB], [QP] and [XY] are parallel and equal in magnitude. It is true to say that:
    [A] **AB** = **XY**   [B] **PQ** = **XY**   [C] **AB** = **PQ**   [D] **AB** = **YX**

15. Two vectors are equal if and only if they have the same:
    [A] Magnitude and direction.   [B] Magnitude and are parallel.
    [C] Magnitude and sense.   [D] Sense and are parallel $\frac{1}{3}(\mathbf{a}+2\mathbf{b})$

Topic 6

TRIGONOMETRY

Objectives
At the end of this topic, the learner should be able to:

1. Appreciate the meaning of trigonometry.
2. State and use the Pythagoras theorem to solve right-angled triangles.
3. Identify Pythagorean triples.
4. Find and use the trigonometric ratios of a right-angled triangle.
5. Find trigonometric ratios and inverse trigonometric ratios using calculators.
6. Find other trigonometric ratios given another.
7. State the complement of an angle and find the trigonometric ratio of complementary angles.
8. Define and memorize trigonometric ratios of the special angles $0°$, $30°$, $45°$, $60°$ and $90°$.

6.1 Meaning of Trigonometry

The word trigonometry comes from three Latin words:
'Tri' means 'three'
'gon' means 'angle'
'metry' means 'measure'

Therefore trigonometry is the study of the relationship between the angles and sides of triangles or simply the study of the measure of triangles.

6.2 Standard Notation for Triangles

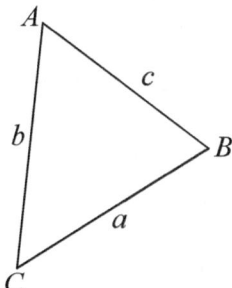

In labelling a triangle, uppercase or capital letters are used to label the vertices (corners) as shown in the figure above. The side opposite a vertex is labelled using its corresponding lowercase or small letter. For instance the side opposite the vertex A is labelled using the letter a. The triangle above is referred to as triangle ABC. The order of the letters does not matter.

6.3 The Right-Angled Triangle

Recall that a **right-angled triangle** is a triangle with one of its angles equal to 90°. The other two angles are each less than 90°. We call an angle, which is equal to 90°, a **right angle**. We often represent a right angle using a small square as shown in the figure below. An **acute angle** is an angle, which is greater than or equal to 0° but less than 90°. In a right-angled triangle, shown in the figure below we call the longest side AC the **hypotenuse** and this side is always opposite to the right angle. We call the two shorter sides AB and BC, which are next to the right angle the **arms** or the **legs** of the right-angled triangle.

Competency Based Mathematics for Secondary Schools. Book 3

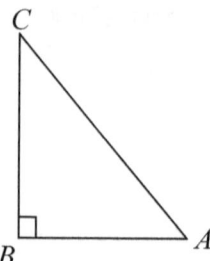

In the figure above, we can reckon the sides a and c in the following ways;
Side a is opposite to angle BAC or a is adjacent or next to angle ACB.
Side c is opposite to angle ACB or c is adjacent or next to angle BAC.

6.4 The Pythagoras Theorem

 Investigative Activity

The figure below shows a right-angled triangle ABC with $AB = 3$ squares and $BC = 4$ squares.
1. Count the number of squares on each arm of the triangle and record your result in the table below.
2. Also count the number of squares on the hypotenuse and record your result in the table below.

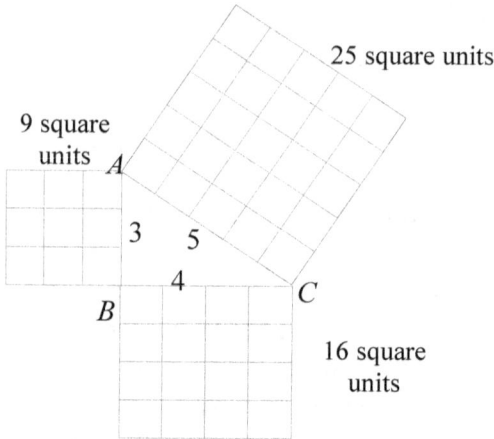

3. On a square paper draw a right-angled triangle ABC right angle at B, with $AB = 5$ squares and $BC = 12$ squares.
4. Cut out part of the square paper and place on the hypotenuse AC of the triangle as in the figure above.
5. What is the length of the hypotenuse in terms of the number of squares?
6. Draw the large squares on each side of the triangle.
7. Count the number of squares on each arm of the triangle and record your result in the table below.

8. Also count the number of squares on the hypotenuse and record your result in the table.
9. Study the table and make a conclusion on your investigation.

Number of squares on side			Sum of squares on AB and BC
AB = c	BC = a	AC = b	

The above investigation leads us to the following theorem called the Pythagoras theorem.

The Pythagoras theorem states that in a right-angled triangle, the number of squares on the hypotenuse is equal to the sum of the squares on the other two sides.

Thus for the triangle in Figure 28:2

$$AC^2 = AB^2 + BC^2$$
$$\Rightarrow AC = \sqrt{AB^2 + BC^2}$$
$$\Rightarrow AB = \sqrt{AC^2 - BC^2}$$
$$\Rightarrow BC = \sqrt{AC^2 + AB^2}$$

To confirm further, this relationship draw right-angled triangles with sides of different lengths, measure their sides and test in the formula.

Test Rule for a Right-Angled Triangle

If in a triangle whose sides are known, the sum of the squares on two of the sides is equal to the square on the other side, the triangle is a right-angle triangle. Otherwise the triangle cannot be a right-angled triangle.

Pythagorean Triples

Any three whole numbers, which can form sides of a right-angled triangle, are called Pythagorean triples. Examples of Pythagorean triples are 3,4,5; 6,8,10; 5,12,13; 8,15,17; 7,24,25 etc

Competency Based Mathematics for Secondary Schools. Book 3

Multiples of Pythagorean Triples

 Investigative Activity

1. Multiplying each of the numbers in the Pythagorean triple 3, 4, 5 by the same integer $n = 1, 2, 3, 4, 5 \ldots$ then record your result in the following table.
2. Calculate $AB^2, BC^2, AC^2, AB^2 + BC^2$ and record your result in the table.

n	AB	BC	AC	AB^2	BC^2	AC^2	$AB^2 + BC^2$
1	3	4	5	9	16	25	25
2	6	8	10	36	64	100	100
3	9	12	15	81	144	225	225
4	12	16	20	144	256	400	400
5	15	20	25	225	400	625	625

3. What conclusion do you draw concerning multiples of Pythagorean triples?

From the table, we see that multiples of the Pythagorean triple 3, 4, 5 are also Pythagorean triples.

Generally, multiples of Pythagorean triples are also Pythagorean triples.

 Exercise 6:1

1. Find the hypotenuse of a right-angled triangle whose arms are 6 cm and 8 cm.
2. One side of a right-angled triangle is 3 cm. If the hypotenuse is 5 cm, find the length of the other side.
3. Find the diagonal of a rectangle whose sides are 40 cm by 9 cm long.
4. The diagonal of a rectangle is 30 cm and one of its sides is 24 cm; find the other side of the rectangle.
5. In a right-angled triangle whose hypotenuse is 20 cm, the ratio of the two arms is 3: 4. Find each arm.
6. Determine which of the following triplets are Pythagorean triples.
   (a) 17, 15, 8   (b) 6, 9, 11   (c) 40, 41, 9   (d) 3, 2, 5   (e) 5, 12, 13
   (f) 7, 9, 12   (g) 14, 17, 20   (h) 6, 8, 10   (i) 3, 6, 8

6.5 Trigonometric Ratios

 Investigative Activity

1. On a square paper draw the right-angled triangles width sides 3, 4, 5 and 6, 8, 10 respectively as shown in the figure below.
2. Use a protractor to measure and confirm that $\angle A = \angle A' = 36.9°$, and $\angle B = \angle B' = 53.1°$.

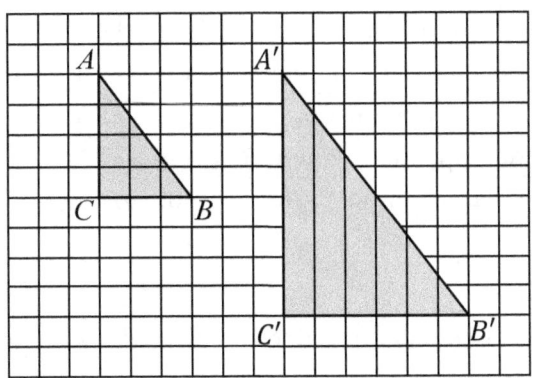

3. Complete the following table to show the ratio of the sides of these triangles.

a	b	c	$\dfrac{a}{c}$	$\dfrac{b}{c}$	$\dfrac{a}{b}$
3	4	5			
6	8	10			

4. What can you deduce about the relationship between the ratios $\dfrac{a}{c}, \dfrac{b}{c}, \dfrac{a}{b}$ and the $\angle A$ and $\angle B$, of any right-angled triangle.

The ratios $\dfrac{a}{c}, \dfrac{b}{c}, \dfrac{a}{b}$ are called **trigonometric ratios** or simply **trig ratios** and depend only on the values of $\angle A$ and $\angle B$, of any right-angled triangle. These ratios are given the special names sine (sin), cosine (cos) and tangent (tan) defined as follows;

Competency Based Mathematics for Secondary Schools. Book 3

$$\sin \hat{A} = \frac{\text{side opposite to angle } A}{\text{hypotenuse}} = \frac{\text{opp}}{\text{hyp}} = \frac{O}{H}$$

$$\cos \hat{A} = \frac{\text{side adjacent to angle } A}{\text{hypotenuse}} = \frac{\text{adj}}{\text{hyp}} = \frac{A}{H}$$

$$\tan \hat{A} = \frac{\text{side opposite to angle } A}{\text{side adjacent to angle } A} = \frac{\text{opp}}{\text{adj}} = \frac{O}{A}$$

Similarly;

$$\sin \hat{B} = \frac{\text{side opposite to angle } B}{\text{hypotenuse}} = \frac{\text{opp}}{\text{hyp}} = \frac{O}{H}$$

$$\cos \hat{B} = \frac{\text{side adjacent to angle } B}{\text{hypotenuse}} = \frac{\text{adj}}{\text{hyp}} = \frac{A}{H}$$

$$\tan \hat{B} = \frac{\text{side opposite to angle } B}{\text{side adjacent to angle } B} = \frac{\text{opp}}{\text{adj}} = \frac{O}{A}$$

Remember that!!
Adjacent to ∠A' means 'next to ∠A' and 'opposite to ∠A' means 'on the other side of ∠A'.

The mnemonic **RAT-SOH-CAH-TOA** may help to commit the definitions to mind. Thus

RAT – means in any Right – Angled Triangle

SOH – stands for $S = \dfrac{O}{H}$ i.e. $\sin = \dfrac{\text{Opposite}}{\text{Hypotenus}}$

CAH – stands for $C = \dfrac{A}{H}$ i.e. $\cos = \dfrac{\text{Adjacent}}{\text{Hypotenus}}$

TOA – stands for $T = \dfrac{O}{A}$ i.e. $\tan = \dfrac{\text{Opposite}}{\text{Adjacent}}$

 Example

Use the figure below to write down as fractions the value of the given trigonometric ratios.
(a) $\sin A$ (b) $\cos C$ (c) $\sin C$ (d) $\cos A$ (e) $\tan A$ (f) $\tan B$

Module 11, Topic 6: Trigonometry

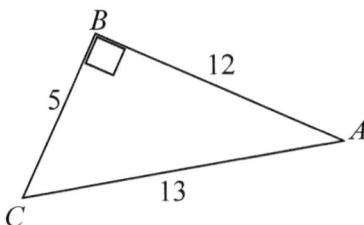

Solution
(a) $\sin A = \dfrac{\text{opp}}{\text{hyp}} = \dfrac{5}{13}$ (b) $\cos C = \dfrac{\text{adj}}{\text{hyp}} = \dfrac{5}{13}$ (c) $\sin C = \dfrac{\text{opp}}{\text{hyp}} = \dfrac{12}{13}$

(d) $\cos A = \dfrac{\text{adj}}{\text{hyp}} = \dfrac{12}{13}$ (e) $\tan A = \dfrac{\text{opp}}{\text{adj}} = \dfrac{5}{12}$ (f) $\tan B = \dfrac{\text{opp}}{\text{adj}} = \dfrac{12}{5}$

 Exercise 6:2

1. In figure (i) and (ii) below, find
    (a) $\tan x$ (b) $\sin x$ (c) $\cos x$ (d) $\tan y$ (e) $\sin y$ (f) $\cos y$

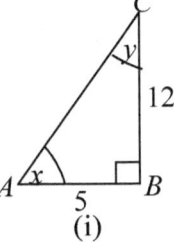 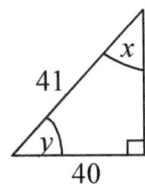
(i) (ii)

2. In the following figure, find the value of
    (a) $\tan x$ (b) $\sin x$ (c) $\cos x$ (d) $\tan y$ (e) $\sin y$ (f) $\cos y$

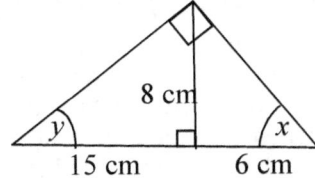

3. If α is acute and $\tan \alpha = \dfrac{7}{24}$, find the value of: (a) $\sin \alpha$ (b) $\cos \alpha$

4. Using figure (i) below, write down the values of:
    (i) $\sin \theta$ (ii) $\cos \theta$ (iii) $\tan \theta$

103

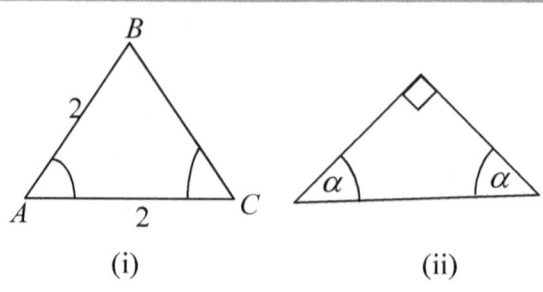

(i) (ii)

5. Using figure (ii) above, write down the value of:
   (i) sin α   (ii) cos α   (iii) tan α

6. Given that x is an acute angle and that $\sin x = \dfrac{m}{n}$, find cos x and tan x in terms of m and n.

7. If α is acute and $\cos A = \dfrac{4}{5}$, find the value of
   (a) sin A (b) tan A

8. In the figure below, find
   (a) cos w (b) sin w (c) tan w (d) cos x (e) sin x (f) tan x
   (g) cos y (h) sin y (i) tan y (j) cos z (k) sin z (l) tan z

6.6 Trigonometric Ratios from Calculators

There are two main ways of finding trigonometric ratios of an angle using scientific calculators depending on the type of calculator. For the purpose of reference, we shall refer to the calculators as Type 1 and Type 2. For Type 1, first key the value of the angle followed by the required trigonometric ratio function key. For Type 2, first key the trigonometric ratio function key followed by the value of the angle. In both cases, first set your calculator to the degree (DEG) mode.

 Example

Use a scientific calculator to find the following:
(a) $\sin 30°$ (b) $\cos 45°$ (c) $\tan 60°$

Module 11, Topic 6: Trigonometry

Solution

	Trig ratio	Type 1	Type 2	Display
(a)	sin 30°	[30] [sin]	[sin] [30]	0.5
(b)	cos 45°	[45] [cos]	[cos] [30]	0.707106...
(c)	tan 60°	[60] [tan]	[tan] [30]	1.73205...

If an angle is in degrees and minutes, it is necessary to convert first the minute part to a decimal of a degree by dividing by 60.

Example

Use a calculator to find the following
(a) sin 47°26'   (b) cos 53°47'   (c) tan 72°13'

Solution
(a) sin 47°26'

Calculator	Procedure	Display
Type 1	[26] [÷] [60] [+] [47] [=] [sin]	0.7364...
Type 2	[sin] [(] [26] [÷] [60] [+] [47] [)] [=]	0.7364...

(b) cos 53°47'

Calculator	Procedure	Display
Type 1	[47] [÷] [60] [+] [53] [=] [cos]	0.5908...
Type 2	[cos] [(] [47] [÷] [60] [+] [53] [)] [=]	0.5908...

(c) tan 72°13'

Calculator	Procedure	Display
Type 1	[13] [÷] [60] [+] [72] [=] [tan]	3.117...
Type 2	[tan] [(] [13] [÷] [60] [+] [72] [)] [=]	3.117...

Competency Based Mathematics for Secondary Schools. Book 3

To find the angle given the trigonometric ratio, the procedures for the two types of calculators are as follows. For Type 1, first punch the value of the trig ratio, followed by the 2nd F (or INV or SHIFT) button followed by the trig function key in that order. For Type 2, first key the 2nd F (or INV or SHIFT) followed by the trig function key and then the value of the angle in that order. Remember to set your calculator to the degree (DEG) mode.

Example

Use a calculator to find the following.
(a) $\sin^{-1} 0.7153$ (b) $\cos^{-1} 0.3431$ (c) $\tan^{-1} 1.34$

Solution

(a) $\sin^{-1} 0.7153$

Calculator	Procedure	Display
Type 1	$\sin^{-1}$ [0.7153] [2nF] [sin]	45.6677...
Type 2	$\sin^{-1}$ [2ndF] [sin] [0.7153] [=]	45.6677...

(b) $\cos^{-1} 0.3431$

Calculator	Procedure	Display
Type 1	$\cos^{-1}$ [0.3431] [2nF] [cos]	69.934...
Type 2	$\cos^{-1}$ [2ndF] [cos] [0.3431] [=]	69.934...

(c) $\tan^{-1} 1.34$

Calculator	Procedure	Display
Type 1	$\tan^{-1}$ [1.34] [2nF] [tan]	53.267...
Type 2	$\tan^{-1}$ [2ndF] [tan] [1.34] [=]	53.267...

Module 11, Topic 6: *Trigonometry*

 Exercise 6:3

1. Use a calculator to find the following trigonometric ratios
   (a) sin 34° (b) sin 72°18' (c) sin 81°20' (d) cos 63° (e) cos 54°36'
   (f) cos 63°43' (g) tan 46° (h) tan 27°6' (i) tan 82°16'
2. Use a calculator to find the angles whose sines are given below.
   (a) 0.1564 (b) 0.9135 (c) 0.9880 (d) 0.8020
   (e) 0.9814 (f) 0.7395 (g) 0.0500 (h) 0.2700
3. Use a calculator to find the angles whose cosines are given below.
   (a) 0.9135 (b) 0.3420 (c) 0.9673 (d) 0.4289
   (e) 0.9586 (f) 0.0084 (g) 0.2611 (h) 0.4700
4. Use calculators to find the angles whose tangent are given below.
   (a) 0.4452 (b) 3.2709 (c) 0.0769 (d) 0.3977
   (e) 0.3568 (f) 1.9251 (g) 0.0163 (h) 0.8263

6.7   Acute Angle Trigonometric Ratios

It has already been seen that the trigonometric ratios are the same for any given angle, no matter the size of the triangle.

 Investigative Activity

1. Use a calculator to complete the following table for integral values of θ for some angles from 0° to 90°.

θ	sin θ	cos θ	tan θ
0	0.0000	1.0000	0.0000
10			
20			
30			
40			
45			
50			
60			
70			
80			
90	1.0000	0.0000	∞

2. Use the table to deduce how the angle θ varies with each of the trigonometric ratios sin θ, cos θ and tan θ.

From the above investigation, we can see that both sin θ and tan θ increase as θ increases, while cos θ decreases as θ increases. Thus;

As θ → 90°, sin θ → 1, tan θ → ∞ and cos θ → 0
As θ → 0°, sin θ → 0, tan θ → 0 and cos θ → 1

Therefore, sin θ and tan θ are increasing functions, while cos θ is a decreasing function.

6.8 Trigonometric Ratios from Tables

We can obtain trigonometric ratios of angles from 0° to 90° from four figure tables. To do this, always make sure that the heading is 'Natural sines' when referring sines, 'Natural cosines' when referring cosines or 'Natural tangents' when referring tangents.

(i) For an angle with an exact number of degrees, look under the column headed 0'.
(ii) For an angle with multiple of 6 as the minutes, look under the column headed by that multiple of 6. If the minutes are not a multiple of 6 subtract the lower but closest multiple of 6 to obtain the difference heading. The difference is added for sine and tangent, because these are increasing functions but subtracted for cosine because it is a decreasing function.

 Example

1. Use tables to find (i) sin 16° (ii) sin 16°24' (iii) sin 16°29'

Solution
Make sure the heading of your table is 'Natural sines'.
(i) Look for sin 16° under 0'. This gives 0.2756.
(ii) Look for sin 16° under 24'. This gives 0.2823.
(iii) Add the difference 14 found under the difference column 5' to the result in (ii) above. Thus
0.2823 + 14 = 0.2837.

Notice that the difference 14 is actually an abbreviation for 0.0014.

2. Use tables to find (i) tan 16° (ii) tan 16°24' (iii) tan 16°29'

Solution
Make sure the heading of your table is 'Natural tangents'. The procedure is as in 1 above.
(i) tan 16° = 0.2867   (ii) tan 16°24' = 0.2943   (iii) tan 16°29' = 0.2959

3. Use tables to find (i) cos 16° (ii) cos 16°24' (iii) cos 16°29'

Solution
Make sure the heading of your table is 'Natural cosines'.
The procedure is as in example 28.2 and 28.3 but for the fact that the difference is subtracted. Thus,
(i) cos 16° = 0.9613   (ii) cos 16°24' = 0.9593   (iii) cos 16°29' = 0.9589.

Note!
Some tables have the natural sines and cosines on one page. On such tables, the angles for cosines are usually on the right most columns, written in descending order and the minute row for cosines at the base of the table also written in descending order.

6.9   Inverse Trigonometric Ratios

Given a trigonometric ratio, the corresponding angle can equally be found. To do this from four figure tables, refer this trigonometric ratio from the body of the table and copy the corresponding angle. If the trigonometric ratio is not found, copy out the angle corresponding to the trigonometric ratio just lower than the one given. Subtract the trigonometric ratio copied from the one given and look for the difference under the remnant. We add the difference in the case of sine and tangent and subtract the difference in the case of cosine. We call the corresponding angle to a given trigonometric ratio the **arc trigonometric ratio** of the angle. For instance, $30° = \text{arc sin } 0.5$. Another way of denoting arc sin 0.5 is $\sin^{-1} 0.5$.

 Example

1. Use tables to find (i) $\sin^{-1} 0.2756$ (ii) $\sin^{-1} 0.2823$ (iii) $\sin^{-1} 0.2837$

Solution
(i) Look for 0.2756 in the body of the table. This would be found under 16° 0'. Therefore, arcsine 0.2756 is 16°.
(ii) Look for 0.2823. This would be found under 16° 24'. Therefore arcsine 0.2756 is 16° 24'.
(iii) 0.2837 is not seen but the next smaller value 0.2823 is seen under 16° 24'. The difference between them is 14 found under 5'. Add this 5' to 24'. Therefore, $\sin^{-1} 0.2837$ is 16°29'.

2. Use tables to find (i) $\tan^{-1} 0.2867$ (ii) $\tan^{-1} 0.2943$ (iii) $\tan^{-1} 0.2959$

Solution
The procedure is the same as in Example 28:5.

(i) $\tan^{-1} 0.2867 = 16°$ (ii) $\tan^{-1} 0.2943 = 16°24'$ (iii) $\tan^{-1} 0.2959 = 16°29'$

3. Use tables to find (i) $\cos^{-1} 0.9613$ (ii) $\cos^{-1} 0.9593$ (iii) $\cos^{-1} 0.9593$

Solution
The procedure is the same as in example 28.5 and 28.6 but for the fact that the difference is subtracted. Thus
(i) $\cos^{-1} 0.9613 = 16°$ (ii) $\cos^{-1} 0.9593 = 16°24'$ (iii) $\cos^{-1}(0.9589) = 16°29'$

Exercise 6:4

1. Use a calculator to find the following trigonometric ratios
   (a) $\sin 34°$ (b) $\sin 72°18'$ (c) $\sin 81°20'$ (d) $\cos 63°$ (e) $\cos 54°36'$
   (f) $\cos 63°43'$ (g) $\tan 46°$ (h) $\tan 27°6'$ (i) $\tan 82°16'$
2. Use a calculator to find the angles whose sines are given below.
   (a) 0.1564 (b) 0.9135 (c) 0.9880 (d) 0.8020
   (e) 0.9814 (f) 0.7395 (g) 0.0500 (h) 0.2700
3. Use a calculator to find the angles whose cosines are given below.
   (a) 0.9135 (b) 0.3420 (c) 0.9673 (d) 0.4289
   (e) 0.9586 (f) 0.0084 (g) 0.2611 (h) 0.4700
4. Use calculators to find the angles whose tangent are given below.
   (a) 0.4452 (b) 3.2709 (c) 0.0769 (d) 0.3977
   (e) 0.3568 (f) 1.9251 (g) 0.0163 (h) 0.8263

6.10 Finding other Trig Ratios Given Another

Example

1. Given that $\cos\theta = \dfrac{15}{17}$, find the values of $\sin\theta$ and $\tan\theta$ without using tables or calculators.

Solution

Using the Pythagoras theorem in the figure below, $BC = \sqrt{AC^2 - AB^2}$
$\Rightarrow BC = \sqrt{17^2 - 15^2} = \sqrt{64} = 8$ units
$\Rightarrow \sin\theta = \dfrac{BC}{AC} = \dfrac{8}{17}$ and $\tan\theta = \dfrac{BC}{AB} = \dfrac{8}{15}$

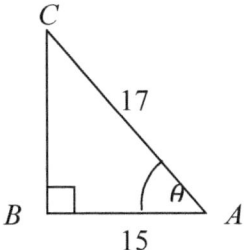

2. Given that $\sin\theta = 0.6$, find the values of $\cos\theta$ and $\tan\theta$ without using tables or calculators.

   Solution

   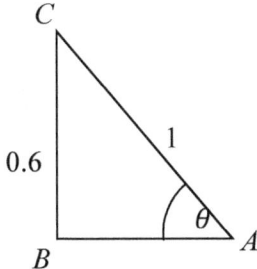

   Using the Pythagoras theorem in the figure above,
   $AB = \sqrt{AC^2 - BC^2}$
   $\Rightarrow AB = \sqrt{1^2 - 0.6^2} = \sqrt{0.64} = 0.8$ units
   $\cos\theta = \dfrac{AB}{AC} = \dfrac{0.8}{1} = 0.8$ and $\tan\theta = \dfrac{BC}{AB} = \dfrac{0.6}{0.8} = 0.75$

 Exercise 6:4

1. If $\cos\alpha = 0.64$, find the values of $\sin\alpha$ and $\tan\alpha$ without using tables or calculators.

2. Given that $\sin x = \dfrac{5}{13}$. Find $\cos x$ and $\tan x$ without using tables or calculators.

3. If $\tan A = \dfrac{4}{3}$, find the values of $\sin A$ and $\cos A$ without using tables or calculators.

6.11 Complementary Angles

Investigative Activity

1. Using square paper, draw 4 different right-angled triangles of different sizes and shapes and label 1, 2, 3 and 4.
2. Label the vertices A, B and C as shown below.

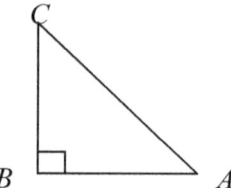

3. Use a protractor to measure the angles A and C.
4. Record your result in the following table.
5. Use a calculator to punch $\sin A$, $\sin C$, $\cos A$ and $\cos C$.
6. Record your result in the table.

	∠A	∠C	∠A + ∠C	sin A	sin C	cos A	cos C
1.							
2.							
3.							
4.							

Use your table to answer the following questions.

7. What conclusion do you draw about the sum of the acute angles of a right-angled triangle?
8. What conclusion do you draw about the sine and cosine of the acute angles of a right-angled triangle?

From the investigation, we see that the sum of the acute angles of a right-angled triangle is 90°.

Two angles whose sum is 90° are called **complementary angles** and one is said to be the complement of the other.

Generally,
> *The sine of any acute angle is equal to the cosine of its complement and vice versa.*

Symbolically, this is expressed as,

$$\left. \begin{array}{l} \sin\theta = \cos(90° - \theta) \\ \cos\theta = \sin(90° - \theta) \end{array} \right\}, \quad 0 \leq \theta \leq 90°$$

Module 11, Topic 6: Trigonometry

Example

Two angles x and y are complementary. Find the value of x if $y = 60°$.

Solution

$$x = 90° - y \text{ and } y = 60°$$
$$\Rightarrow x = 90° - 60° = 30°$$

Exercise 6:5

1. Given that x, y and z are angles of a right-angled triangle. Find the missing value.

	$x°$	$y°$	$z°$
(i)	72	90	—
(ii)	25	—	90
(iii)	43.5	—	90
(iv)	—	57	33
(v)	90	22.5	—
(vi)	—	13	77

2. From the following figure, find $\cos \emptyset$.

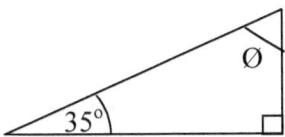

3. Write down the value of the missing angle in figure (a) to (d)

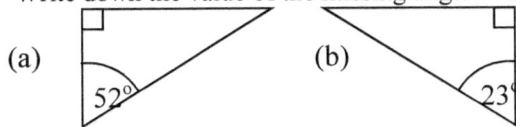

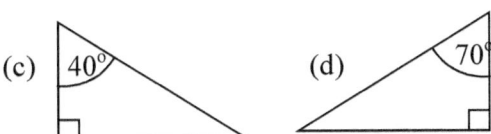

4. State the complement of each of the following angles.
    (a) 60°   (b) 34°   (c) 45°   (d) 90°   (e) 0°
5. Given that $\sin 48° = 0.7431$. Find the value of $\cos 42°$ without using tables or calculator.

6. Given that cos 63 = 0.4540, without using tables or calculator find the value of sin 27°.
7. In the following triangle, write down 20 possible pairs of values of x and y.

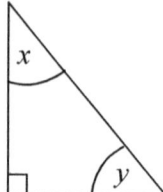

8. State two possible angles of a right-angled triangle and draw the triangle marking the value of each of its angles.

6.12 Trig Ratios of Special Angles

The trigonometric ratios of certain angles are used so frequently that they need to be given special attention. These are the trigonometric ratios of 0°, 30°, 45°, 60° and 90°.

Trigonometric Ratios of 30° and 60°

 Investigative Activity

1. Draw an equilateral triangle ABC with each side equal to 2 units
2. Draw a perpendicular BN from the vertex B to a point N on the opposite side AC of the triangle.

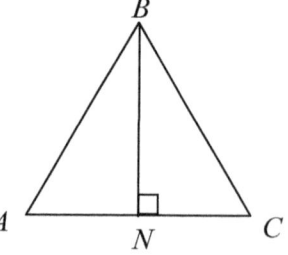

3. State the value of each of the angles $\angle ABC, \angle BAC, \angle BCA, \angle ABN, \angle CBN$.
4. Find the length of the sides AN, NC and BN.
5. What type of triangle is each of $\triangle ABN$ and $\triangle CBN$.
6. Use the Pythagoras theorem, the definition of trigonometric ratios and your facts in 3, 4 and 5 to find the values of sin 30°, cos 30°, tan 30°, sin 60°, cos 60°, tan 60°.

Module 11, Topic 6: Trigonometry

From the above investigation, we see that;

$\sin 60° = \dfrac{\sqrt{3}}{2}$	$\cos 30° = \dfrac{\sqrt{3}}{2}$	$\tan 60 = \dfrac{1}{\sqrt{3}} = \dfrac{\sqrt{3}}{3}$
$\sin 30° = \dfrac{1}{2}$	$\cos 60° = \dfrac{1}{2}$	$\tan 30° = \dfrac{\sqrt{3}}{1} = \sqrt{3}$

Trigonometric Ratios of $45°$

 Investigative Activity

1. Draw a right-angled isosceles triangle ABC with each arm equal to 1 unit.

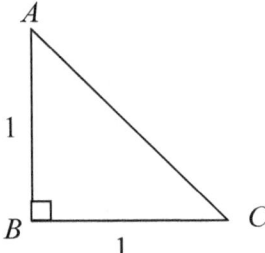

2. State the value of each of the angles $\angle ABC, \angle BAC, \angle BCA$.
3. Find the length of the side AC.
4. Use the Pythagoras theorem, the definition of trigonometric ratios and your facts in 1, 2 and 3 to find the values of $\sin 45°, \cos 45°, \tan 45°$.

From the above it is clear that $\angle ABC = 90°$ and $\angle BAC = \angle BCA = 45°$. Using the Pythagoras theorem, we can calculate the hypotenuse AC of triangle ABC. Thus;

$$AC = \sqrt{1^2 + 1^2} = \sqrt{2}$$

Therefore the trigonometric ratios of $45°$ are as follows.

$\sin 45° = \dfrac{1}{\sqrt{2}} = \dfrac{\sqrt{2}}{2}$	$\cos 45° = \dfrac{1}{\sqrt{2}} = \dfrac{\sqrt{2}}{2}$	$\tan 45 = \dfrac{1}{1} = 1$

Trigonometric Ratios of $0°$ *and* $90°$

Consider the right-angled triangle ABC above with angle $BAC = \theta$ and angle $BCA = \alpha$. Suppose that C is a moving point along BQ, then as C moves along BQ, the line AC rotates about A as centre. Thus, we can make the angle as small as possible (or nearly 0°). At the same time as θ is getting smaller and smaller, α is getting larger and larger and approaching the value of 90°. Under these conditions, the length AC approximates the length AB and the length BC approximates 0.

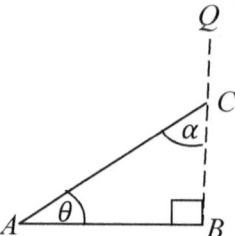

Symbolically;
as $\theta \to 0°$, $\alpha \to 90°$, $AC \to AB$ and $BC \to 0$

Since $\sin\theta = \dfrac{BC}{AC}$, $\qquad \operatorname{cosec}\theta = \dfrac{AC}{BC}$,

$\cos\theta = \dfrac{AB}{AC}$, $\qquad \sec\theta = \dfrac{AC}{AB}$,

$\tan\theta = \dfrac{BC}{AB}$, $\qquad \tan\theta = \dfrac{AC}{BC}$

also $\sin\alpha = \dfrac{AB}{AC}$, $\qquad \operatorname{cosec}\alpha = \dfrac{AC}{AB}$,

$\cos\alpha = \dfrac{BC}{AC}$, $\qquad \sec\alpha = \dfrac{AC}{BC}$,

$\tan\alpha = \dfrac{AB}{BC}$, $\qquad \tan\alpha = \dfrac{BC}{AB}$

The limiting values are $\theta = 0°$, $\alpha = 90°$, $AB = AC$ and $BC = 0$.

Substituting these limiting values,

$\sin 0° = \dfrac{0}{AB} = 0$, $\qquad \operatorname{cosec} 0° = \dfrac{AB}{0} = \infty$,

$\cos 0° = \dfrac{AB}{AB} = 1$, $\qquad \sec 0° = \dfrac{AB}{AB} = 1$,

$\tan 0° = \dfrac{0}{AB} = 0$, $\qquad \tan 0° = \dfrac{AB}{0} = \infty$

$\sin 90° = \dfrac{AB}{AB} = 1$, $\qquad \operatorname{cosec} 90° = \dfrac{AB}{AB} = 1$,

Module 11, Topic 6: Trigonometry

$$\cos 90° = \frac{0}{AB} = 0, \qquad \sec 90° = \frac{AB}{0} = \infty,$$

$$\tan 90° = \frac{AB}{0} = \infty, \qquad \tan 90° = \frac{0}{AB} = 0$$

We can summarize these results as in the following table.

θ	$\sin\theta$	$\cos\theta$	$\tan\theta$	$\csc\theta$	$\sec\theta$	$\cot\theta$
0°	0	1	0	∞	1	∞
30°	$\frac{1}{2}$	$\frac{\sqrt{3}}{2}$	$\frac{\sqrt{3}}{3}$	2	$\frac{2\sqrt{3}}{3}$	$\sqrt{3}$
45°	$\frac{\sqrt{2}}{2}$	$\frac{\sqrt{2}}{2}$	1	$\sqrt{2}$	$\sqrt{2}$	1
60°	$\frac{\sqrt{3}}{2}$	$\frac{1}{2}$	$\sqrt{3}$	$\frac{2\sqrt{3}}{3}$	2	$\frac{\sqrt{3}}{3}$
90°	1	0	∞	0	∞	0

Exercise 6:7

1. Compute the following without using tables or calculators. Where appropriate leave your answer in surd form.
   (a) $\cos^2 60° + \tan^2 45°$ (b) $\sin^2 30° - \cos^2 60°$ (c) $2 - \cos 60°$
   (e) $1 + \tan^2 30°$ (f) $\sin^2 30° + \cos^2 30°$
2. Without tables or calculators, evaluate the following leaving your answers in surd form.   (a) $\cos 60° + \sin 30°$ (b) $\sin 60° + \cos 30°$

6.13 Real Life Application of Trigonometry

Trigonometry is very useful in solving many real life geometric problems.

Example

1. Two opposite sides of a cage are 40 cm apart and one side is 9 cm taller than the other. Calculate the length of its slanting roof and the angle between the roof and the horizontal.

Competency Based Mathematics for Secondary Schools. Book 3

Solution

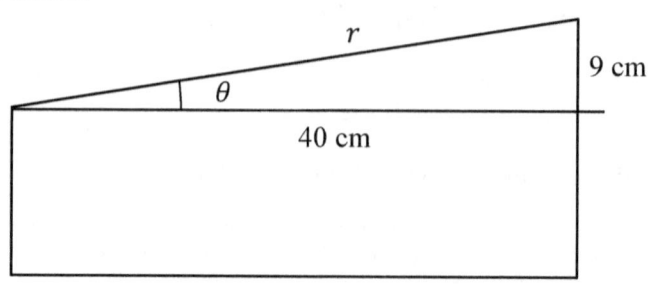

Using the Pythagoras theorem, $r = \sqrt{40^2 + 9^2} = 41$ cm.
We can use any of the three trig ratios to find θ.
Thus, $\sin \theta = \frac{9}{41} \Rightarrow \theta = \sin^{-1}\left(\frac{9}{41}\right) = 12.7°$

2. A carpenter places a ladder against the wall of a 4 m tall building. The angle between the upper edge of the ladder and the wall is 60°. Determine,
   (a) the length of the ladder
   (b) the distance of the ladder from the wall to two decimal places.
   (c) the angle between the ladder and the ground.

Solution
Let the length of the ladder be l and the distance from the wall be x.

(a) $\frac{4}{l} = \cos 60$

$\Rightarrow \frac{4}{l} = \frac{1}{2}$

Cross multiplying we have
$l = 8$ m

Length of the ladder = 8 m

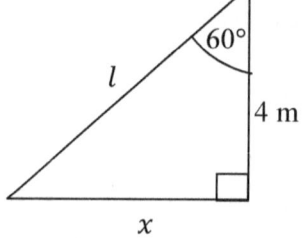

(b) $\frac{x}{4} = \tan 60° \Rightarrow x = 4 \tan 60° = 4\sqrt{3} \approx 6.93$

Distance from the wall $x \approx\approx 6.93$ m

Alternatively, we can use the Pythagoras theorem to find the side x

Thus, $x = \sqrt{8^2 - 4^2} = \sqrt{48} \approx 6.93$ m.

Exercise 6:8

1. A ladder leans against a vertical wall and makes an angle of 70° with the horizontal ground. The foot of the ladder is 3 m from the wall. How far up the wall to the nearest m does the ladder reach?
2. The figure below shows the cross-section of a roof. How long must the carpenter cut the plank support *BN*? What is the length of the base *AC*?

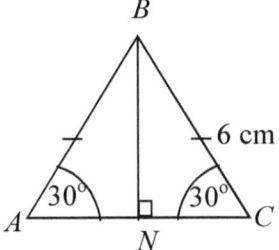

3. A road rises 105 m for every *x* m of the horizontal. If the angle between the slope and the horizontal is 6°, find the value of *x*.
4. The roof of a bamboo hurt is a triangular prism. If the altitude of the roof from the ceiling is, 3 m and the equal slanting sides make each an angle of 30° with the horizontal. Calculate:
   (a) The length of the slanting sides.
   (b) The angle made by the slanting sides at the upper edge.
   (c) The length of the base between the two slanting sides.
5. A plane takes off at *A* and ascends at a fixed angle of 22° with the level ground. After it flies 3000 m, find to the nearest metre;
   (a) The altitude of the plane   (b) The distance covered horizontally.

 Multiple Choice Exercise 6

1. A triangle has sides 8 cm, 15 cm and 17 cm. Therefore, the best name for it is:
   [A] a equilateral triangle.   [B] an obtuse triangle.
   [C] a right-angled triangle.   [D] an isosceles triangle.
2. The pair of trigonometric ratios, which are equal, is:
   [A] sin 50° and cos 50°   [B] sin 50° and tan 50°
   [C] sin 50° and tan 40°   [D] sin 50° and cos 40°
3. Given the following triangle, the incorrect relation is:
   [A] $r^2 = p^2 + q^2$ [B] $p^2 = r^2 - q^2$ [C] $q^2 = r^2 - p^2$ [D] $q^2 = r^2 + p^2$

Competency Based Mathematics for Secondary Schools. Book 3

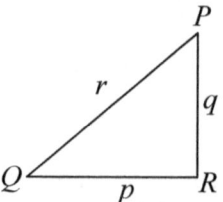

4. Given the triangle above, the value of p is:
   [A] $\sqrt{q^2 - r^2}$ [B] $\sqrt{r^2 - q^2}$ [C] $\sqrt{q^2 + r^2}$ [D] $r^2 + q^2$
5. A Pythagorean triple among the following is:
   [A] 6, 9, 11   [B] 7, 9, 12   [C] 8, 15, 17   [D] 14, 17, 20
6. The triplet, which does not represent the lengths of the sides of a right-angled triangle is:
   [A] 6, 8, 10   [B] 5, 12, 10   [C] 8, 15, 17   [D] 7, 23, 24
7. 36.4° is equal to:
   [A] 36°34'   [B] 36°60'   [C] 36°24'   [D] 36°40'
8. 41°27' in degrees as a decimal is:
   [A] 41.45°   [B] 41.33°   [C] 41.25°   [D] 41.60°
9. The ratio, which is equal to sin 60°, is:
   [A] cos 30°   [B] sin 30°   [C] cos 60°   [D] tan 30°
10. The ratio, which is equal to sin 30°, is:
    [A] 0.0500   [B] 0.5050   [C] 0.866   [D] 0.5000
11. The value of tan 45° is:
    [A] 2.0000   [B] 0.5000   [C] 1.0000   [D] 1.5000
12. The cosine of 60° is equal to $\frac{1}{2}$. The value of 28–20cos 60° is:
    [A] 18   [B] 10   [C] 4   [D] 24
13. If $\sin \theta = \frac{3}{5}$, the value of $\tan \theta$ for 0<θ<90° is:
    [A] $\frac{4}{5}$ [B] $\frac{3}{4}$ [C] $\frac{5}{8}$ [D] $\frac{1}{2}$
14. If $\cos \theta = \frac{5}{13}$ the value of $\tan \theta$ for 0<θ<90° is:
    [A] $\frac{5}{12}$ [B] 5   [C] $\frac{13}{5}$ [D] $\frac{12}{5}$
15. Given that $\tan x = \frac{5}{12}$. The value of sin x + cos x is:
    [A] $\frac{5}{13}$ [B] $\frac{7}{13}$ [C] $\frac{17}{13}$ [D] $\frac{5}{12}$
16. If $\tan x = 2\frac{1}{2}$ the value of sin x for 0°<x<90° is:

[A] $\frac{\sqrt{29}}{5}$ [B] $\frac{5\sqrt{29}}{29}$ [C] $\frac{2\sqrt{29}}{29}$ [D] $\frac{\sqrt{29}}{2}$

17. If $\sin x = \cos 50°$ then, x equals:
 [A] 40° [B] 45° [C] 50° [D] 90°
18. If $\sin(x+30)° = \cos 40°$, the value of $x°$ is:
 [A] 15° [B] 20° [C] 60° [D] 90°
19. If $10 \tan 60° = 20 \tan x$. Correct to the nearest degree x is equal to:
 [A] 30° [B] 40° [C] 41° [D] 60°
20. If $\sin x = \cos 70°$, x equals:
 [A] 110° [B] 70° [C] 30° [D] 20°
21. If $\cos x = \sin 27.3°$, the value of x where $0° \leq x \leq 90°$ is:
 [A] 27.3° [B] 35.4° [C] 54.6° [D] 62.7°
22. Without using tables or calculators, the value of $\frac{\sin 20°}{\cos 70°} + \frac{\cos 25°}{\sin 65°}$ is:
 [A] 2 [B] 1 [C] –2 [D] –1
23. If $\sin x = \frac{12}{13}$, where $0° < x < 90°$. The value of $1 - \cos^2 x$ is:
 [A] $\frac{25}{169}$ [B] $\frac{64}{169}$ [C] $\frac{105}{169}$ [D] $\frac{144}{169}$
24. Using four figure tables, the sine of 70° is:
 [A] 0.9390 [B] 0.9394 [C] 0.9397 [D] 0.9399
25. Using four figure tables cos 80° is:
 [A] 0.173 [B] 0.1736 [C] 0.1740 [D] 0.1744
26. Using four figure tables, the angle whose sine is 0.841 to one decimal place is:
 [A] 57.2° [B] 56.8° [C] 32.8° [D] 32.2°
27. Given that $\cos x = 0.5321$, then x is equal to:
 [A] 56.2° [B] 57.9° [C] 33.2° [D] 32.1°
28. Using Mathematical tables $\cos 40° - \sin 30°$ equals:
 [A] – 0.2660 [B] 0.2660 [C] – 0.0266 [D] 0.0266
29. In $\triangle PQR$, $\angle PQR$ is a right angle $|QR| = 2$ cm and $\angle PRQ = 60°$. $|PR|$ is equal to:
 [A] $4\sqrt{3}$ cm [B] 4 cm [C] $2\sqrt{3}$ cm [D] 1 cm
30. 25° 45' as a decimal is:
 [A] 25.75° [B] 25.55° [C] 25.45° [D] 25.15°
31. In surd form $\sin 45° \cos 30° + \cos 45° \sin 30°$ is equal to:
 [A] $\frac{\sqrt{2}}{2}$ [B] $\frac{\sqrt{3}}{2}$ [C] $\sqrt{2}$ [D] $\frac{\sqrt{6}+\sqrt{2}}{4}$
32. In the triangle shown below the length of the side marked x is:
 [A] 7sin 56° [B] 7tan 56° [C] 7tan 34° [D] 7cos 34°

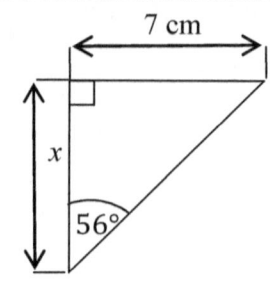

33. In the figure below, $AC = 42$ cm, $AD = 12$ cm and $BD = 16$ cm. As a vulgar fraction tan A is:

   [A] $\dfrac{4}{5}$ [B] $\dfrac{3}{5}$ [C] $\dfrac{5}{6}$ [D] $\dfrac{4}{3}$

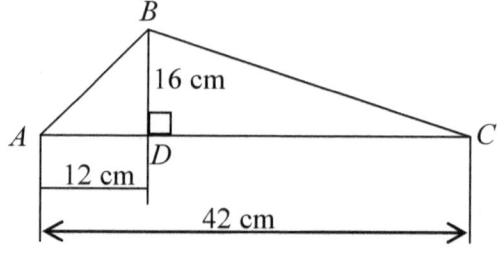

34. In the figure above, $AC = 42$ cm, $AD = 12$ cm and $BD = 16$ cm. Sin A as a decimal fraction is:
   [A] 0.8   [B] 1.3   [C] 1.6   [D] 0.6
35. In the figure above, $AC = 42$ cm, $AD = 12$ cm and $BD = 16$ cm. The size of C to the nearest degree is:
   [A] 27°   [B] 32°   [C] 28°   [D] 58°
36. In the figure above, $AC = 42$ cm, $AD = 12$ cm and $BD = 16$ cm. The perimeter of triangle BDC in cm is:
   [A] 70   [B] 80   [C] 90   [D] 92
37. $ABCD$ is a trapezium. Angle D is a right angle and BE is perpendicular to DC. $AB = 12$ cm and $AD = 8$ cm. The value of $DC = 18$ cm. The value of tan C is:

   [A] $\dfrac{4}{9}$ [B] $\dfrac{2}{3}$ [C] $\dfrac{4}{3}$ [D] $\dfrac{4}{5}$

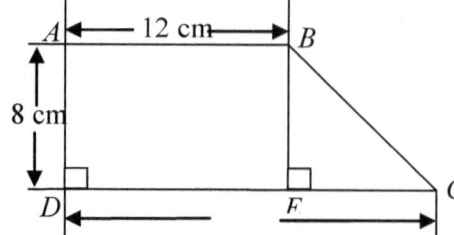

Module 11, Topic 6: Trigonometry

38. A ladder 9 m long leans against a vertical wall, making an angle of 64° with the horizontal ground. To one decimal place, the distance of the foot of the ladder from the wall is:
    [A] 3.9 m   [B] 5.8 m   [C] 7.9 m   [D] 8.1 m

39. A ladder leans against the wall at an angle 60° to the wall. If the foot of the ladder is 5 m away from the wall, the length of the ladder is:
    [A] $\dfrac{5\sqrt{3}}{3}$ m   [B] 5 m   [C] $5\sqrt{3}$ m   [D] $\dfrac{10\sqrt{3}}{3}$ m

40. A ladder 6 m long leans against a vertical wall, so that it makes an angle of 60° with the wall. The distance of the foot of the ladder from the wall is:
    [A] 3 m   [B] 6 m   [C] $2\sqrt{3}$ m   [D] $3\sqrt{3}$ m

41. In the figure below, ∠PQR is a right angle |QR| = 2 cm and ∠PRQ = 60°. |PR| is equal to:
    [A] 1 m   [B] 4 m   [C] $2\sqrt{3}$ m   [D] $\dfrac{4}{3}$ m

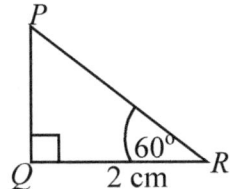

42. A ladder 5 m long rest against a wall so that it foot makes an angle of 30° with the horizontal. The distance of the foot of the ladder from the wall is:
    [A] $\dfrac{5\sqrt{3}}{3}$ m   [B] $2\dfrac{1}{2}$ m   [C] $\dfrac{5\sqrt{3}}{2}$ m   [D] $\dfrac{10\sqrt{3}}{3}$ m

43. A pole of length l leans against a vertical wall so that it makes an angle of 60° with the horizontal ground. If the top of the pole is 8 m above the ground, l must be:
    [A] $16\sqrt{3}$ m   [B] $\dfrac{\sqrt{3}}{16}$ m   [C] 16 m   [D] $\dfrac{16\sqrt{3}}{3}$ m

Module 12

Solid Figures

Family of Situations
Module 12 is an extension of module 3 and 7. At the end of the module; the student is expected to acquire many more competencies within the **families of situations** *'Usage of Technical Objects in everyday life'*.

Categories of Action
The categories of action for module 12 include:
1. Recognition of objects,
2. Production of commodities or provision for daily consumption,
3. production of parts for industrial use,
4. production of materials for work of arts and construction.
5. Determination of measures.

Credit
The module is expected to be covered within 3 weeks teaching 4 periods of 50 minutes per week (or within 10 periods).

Topic 7

MENSURATION OF SOLIDS

Objectives

At the end of this topic, the learner should be able to:

1. Describe and identify a sphere.
2. Give examples of spheres in real life.
3. Draw a sphere.
4. Find the surface area and volume of a sphere.
5. Describe and identify prisms, cones, pyramids, frustums.
6. Give examples of prisms, cones, pyramids, frustums in real life.
7. Draw prisms, cones, pyramids, frustums.
8. Draw and make nets of prisms, cones, pyramids, frustums
9. Find the surface area and volume of prisms, cones, pyramids, frustums.
10. Find the ratio of surface area and volumes of solids.
11. Distinguish between a prism and a pyramid.

For vocabulary associated with all the solid figures treated in this topic, consult module 3 topic 14 and module 7 topics 9 and 10.

7.1 Surface Area and Volume of Prisms

A right prism has a pair of parallel and congruent polygonal base.
Consider a prism of length l, cross-sectional area A and cross-sectional perimeter p.

The surface area S of the prism is given by $\quad S = 2A + pl$ ....①

The volume V of the prism is given by $\quad V = Al$ ....②

Example

1. A right-angled triangular prism has arms 4 cm and 3 cm and length 10 cm. Calculate (a) its surface area. (b) its volume.

   Solution

   $$S = 2A + pl$$
   Hypotenuse of triangle $= \sqrt{3^2 + 4^2} = 5$ cm
   $\Rightarrow p = 3 \text{ cm} + 4 \text{ cm} + 5 \text{ cm} = 12$ cm

   (a) $A = \dfrac{1}{2}bh = \dfrac{1}{2}(4)(3) = 6 \text{ cm}^2$

   $\therefore S = 2(6 \text{ cm}^2) + (12 \text{ cm})(10 \text{ cm}) = 132 \text{ cm}^2$

   (b) $V = Al$, $A = 6 \text{ cm}^2$, $l = 10$ cm
   $\Rightarrow V = 6 \text{ cm}^2 (10 \text{ cm}) = 60 \text{ cm}^3$

Module 12, Topic 7: Mensuration of Solids

7.2 Cones and Pyramids

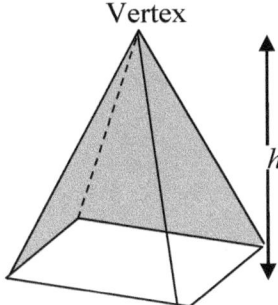
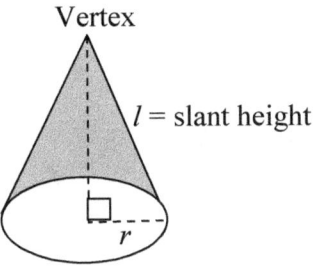

Volume of Cones and Pyramids

The volume V of any cone or pyramid with base area A and vertical height h is $\frac{1}{3}$ that of a cylinder or prism with the same base area and vertical height. Hence,

$$V = \frac{1}{3}\text{base area} \times \text{height}$$
$$V = \frac{1}{3}bh$$

 Example

1. The base of a pyramid is a square of side 5 cm and its height 12 cm. Find its volume.

   Solution
   $$V = \frac{1}{3}Ah = \frac{1}{3}(5 \text{ cm})^2 (12 \text{ cm}) = 100 \text{ cm}^3$$

2. Calculate the volume of a cone whose height is 12 cm and whose base radius is 7 cm.

   Solution
   $$V = \frac{1}{3}Ah = \frac{1}{3}\pi r^2 h \Rightarrow V = \frac{1}{3}\left(\frac{22}{7}\right)(7)^2 (12) \text{ cm}^3 = 616 \text{ cm}^3$$

Surface Area of a Cone

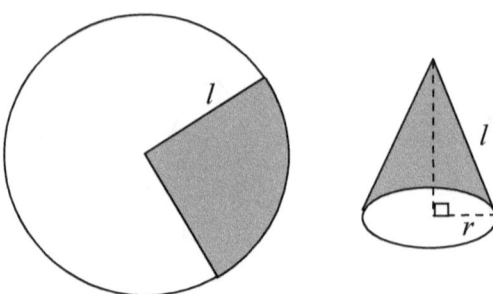

We can make a cone by folding a sector, which subtends an angle θ at the center of a circle. In this way, the radius l of the circle will be the slant height of the cone with radius r. The lateral surface area of the cone will be the area S of the sector. The circumference $C = 2\pi r$ of the base of the cone will be the arc length of the sector.

$$\Rightarrow S = \frac{\theta}{360} \times \pi l^2 \quad \text{.............................①}$$

$$2\pi r = \frac{\theta}{360} \times 2\pi l$$

$$\Rightarrow r = \frac{\theta}{360} \times l \quad \text{.............................②}$$

① ÷ ② : $\quad \dfrac{S}{r} = \pi l$

Therefore, the formula for finding the lateral surface area S of a cone is

$$S = \pi r l$$

Since $l = \sqrt{h^2 + r^2}$, the lateral surface area of a cone is

$$S = \pi r \sqrt{h^2 + r^2}$$

The total surface area of a solid cone will therefore be

$$S = \pi r^2 + \pi r l \quad \text{or} \quad S = \pi r (r + l)$$

In terms of h, the total surface area of a solid cone will be

$$S = \pi r^2 + \pi r \sqrt{h^2 + r^2} \quad \text{or} \quad S = \pi r \left(r + \sqrt{h^2 + r^2} \right)$$

From equation ②, we can see that the radius of the base of a cone, made out of a sector of radius l that subtends an angle θ at the center is

$$r = \frac{\theta l}{360°}$$

 Example

Calculate the surface area of a solid cone with base radius 3.5 cm and slant height 5.5 cm.

Solution
$$S = \pi r (r + l)$$
$$\Rightarrow S = \frac{22}{7}(3.5)(3.5+5.5) = 11(9) \text{ cm}^2 = 99 \text{ cm}^2$$

Surface Area of a Pyramid

To calculate the surface area S of a pyramid, calculate the area of all the faces independently and add to the area of the base. Thus Surface area S of pyramid is given by

S = Area of base + Sum of area of all the triangular faces

 Example

The faces of a pyramid are made of 4 isosceles triangles each with a base 8 cm and height 12 cm. Calculate the surface area of the pyramid.

Solution

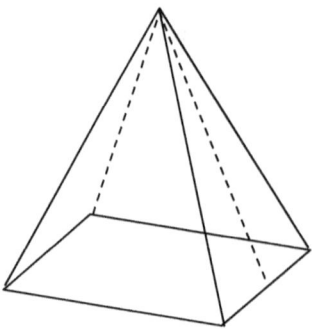 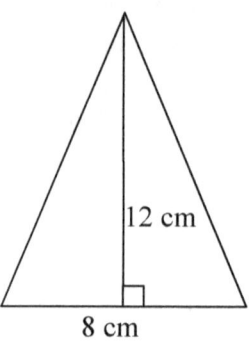

Area of each triangle $= \frac{1}{2}bh = \frac{1}{2}(8)(12) = 48$ cm^2

∴ Area of 4 faces $= 4(48) = 192$ cm^2

Area of base $= (8 \text{ cm})^2 = 64$ cm^2

Surface area $= 64$ cm$^2 + 192$ cm$^2 = 256$ cm^2

Exercise 7:1

1. A cone is 8 cm tall and has a base diameter of 12 cm. Find its slant height.
2. Find the surface area of a solid cone whose base radius is 7 cm and its slant height is 14 cm.
3. Find the surface area of a solid cone of radius 7 cm and slant height 9 cm.
4. The vertical angle of a cone is 70° and its slant height is 11 cm. Calculate the height of the cone correct to the nearest whole number.
5. The radius of a cone is 6 cm. Given that the height of the cone is 14 cm, calculate the volume of the cone.
6. Calculate the area of the surface of a cone with radius 14 cm and slant height 20 cm, when; (a) The base is open   (b) the base is closed.
7. Find the volume of a cone with radius 7 cm and slant height 13 cm
8. A cone has a diameter of 7 cm and a height of 12 cm. Calculate the volume of the cone.
9. Calculate the volume of a cone with base diameter 10 cm and slant height 13 cm.
10. The base of a pyramid is a square. Each face is an isosceles triangle with base 12 cm and height 16 cm. Calculate (a) its surface area (b) its volume
11. Figure (a) below shows a pyramid on a square base. If all the eight edges of the pyramid are equal, find, in degrees, the value of angle OAC.
12. Figure (b) below shows the shaded sector of a circle of radius 21 cm, which subtends an angle of 120° at the center. Find the radius, in metres, of the base of the cone formed from a cone this sector.

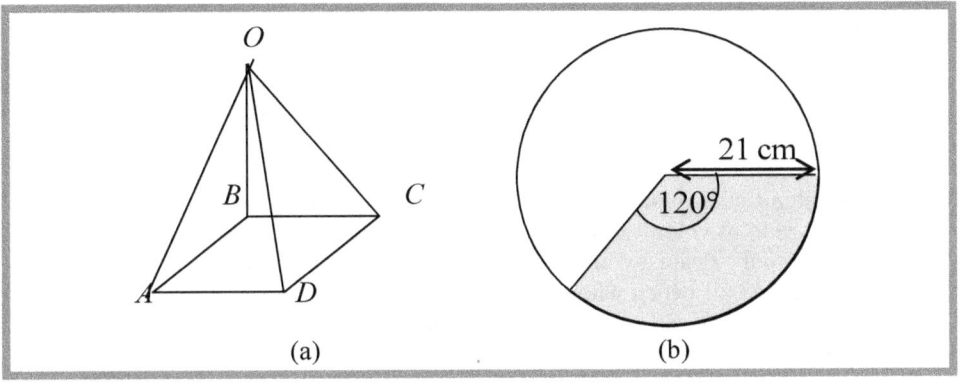

(a)   (b)

7.3 Surface Area and Volume of a Sphere

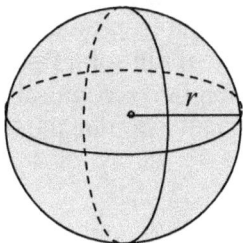

The surface area S of a sphere with radius r is four times the area of a circle with radius r.

Area of sphere, $S = 4 \times$ Area of circle $= 4\pi r^2$

The volume V of the sphere is given by, $V = \dfrac{4}{3}\pi r^3$

 Example

Calculate the surface area and the volume of a sphere whose radius is 14 cm.

Solution

$S = 4\pi r^2 = 4\left(\dfrac{22}{7}\right)(14)^2 = 2464 \text{ cm}^2$

$V = \dfrac{4}{3}\pi r^3 = \dfrac{4}{3}\left(\dfrac{22}{7}\right)(14)^3 = 11498.8 \text{ cm}^3$

Competency Based Mathematics for Secondary Schools. Book 3

 Exercise 7:2

1. Find the surface area and volume of the spheres with the following radius.
   (a) 21 cm   (b) 10.5 cm
2. The surface area of a sphere is 616 cm^2, what is the volume of the sphere correct to two significant figures?
3. Find the ratio of volumes of two spheres whose radii are in the ratio 1:7.
4. Calculate to the nearest whole number the volume of air required to fill completely a ball which when fully expanded has a diameter of 15 cm.
5. A blacksmith melts a lead sphere of diameter 100 mm and uses it to make spherical balls of diameter 1 mm. If no lead is lost in the process, calculate the number of small spheres he makes.
6. A hemisphere has a diameter of 6 cm. Calculate
   (a) Its Volume (b) The area of its curved surface. (c) Its total surface area.

7.4 Composite Solid Figures

As with the case of composite plane figures, problems on solid figures may involve a combination of two or more of the solid figures studied above. Such composite figures must then be broken into recognisable figures. As examples of special composite solid figures, we shall examine the frustum and the hemisphere.

7.5 Volume and Surface Area of a Frustum

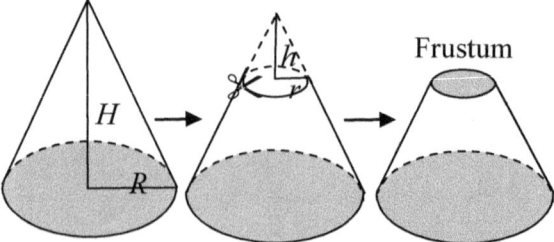

When the smaller end of a cone or pyramid is cut off through its cross-section by a plane parallel to the base as shown in the figure above the portion left is called a **frustum**.

Consider a conical frustum of height h, made by cutting off a small cone of radius r from a larger cone of radius R. The figure below is the diametrical cross-section of this cone.

Module 12, Topic 7: Mensuration of Solids

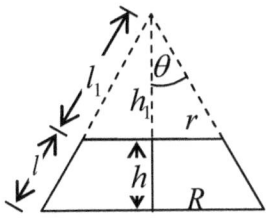

$$\tan\theta = \frac{R}{h+h_1} = \frac{r}{h_1} \Rightarrow h_1 = \frac{rh}{R-r} \quad \text{\textcircled{1}}$$

$$\sin\theta = \frac{R}{l+l_1} = \frac{r}{l_1} \Rightarrow l_1 = \frac{rl}{R-r} \quad \text{\textcircled{2}}$$

Since the volume of a cone with radius r and height h is $V = \frac{1}{3}\pi r^2 h$, the volume of this frustum will be $V = \frac{1}{3}\pi R^2(h+h_1) - \frac{1}{3}\pi r^2 h_1$ ....\textcircled{3}

Substituting \textcircled{1} in \textcircled{3} and rearranging gives

$$V = \frac{1}{3}\pi h\left(R^2 + rR + r^2\right)$$

Since the lateral surface area of a cone with radius r and height h is $A = \pi r l$, the lateral surface area of the frustum will be

$$S_l = \pi R(l+l_1) - \pi r l_1 \quad \text{.........\textcircled{4}}$$

Substituting \textcircled{2} in \textcircled{4} and rearranging gives

$$S_l = \pi(R+r)l$$

But $l = \sqrt{h^2 + (R-r)^2}$

Therefore, $S_l = \pi(R+r)\sqrt{h^2 + (R-r)^2}$

If the frustum is a solid, we add the area of the two circular ends. Hence, surface area of a solid frustum is

$$S = \pi\left(R^2 + r^2 + (R+r)l\right)$$

OR

$$S = \pi\left(R^2 + r^2 + (R+r)\sqrt{h^2 + (R-r)^2}\right)$$

For a frustum open at the larger end such as a bucket the surface area will be

$$S = \pi\left(r^2 + (R+r)l\right)$$

OR

$$S = \pi\left(r^2 + (R+r)\sqrt{h^2 + (R-r)^2}\right)$$

Example

1. A boy cuts cone with base radius 4 cm from a cone of radius 6 cm and height 4 cm. Find
   (a) volume of the frustum left.
   (b) the total surface area of the frustum left.

Solution
Let V, v, R, r and H, h be the volume, radius and height of the big and small cones respectively.

$$\tan\theta = \frac{4}{h} = \frac{6}{4}$$

$$\Rightarrow h = \frac{8}{3} \text{ cm}$$

(a) $V = \frac{1}{3}\pi h\left(R^2 + rR + r^2\right)$

$$= \frac{1}{3}\left(\frac{22}{7}\right)\left(\frac{4}{3}\right)\left(6^2 + 6(4) + 4^2\right)$$

$$= \frac{1}{3}\left(\frac{22}{7}\right)\left(\frac{4}{3}\right)(76) = 106.2 \text{ cm}^3$$

(b) $S = \pi\left(R^2 + r^2 + (R+r)\sqrt{h^2 + (R-r)^2}\right)$

Module 12, Topic 7: Mensuration of Solids

$$= \left(\frac{22}{7}\right)\left(6^2 + 4^2 + (6+4)\sqrt{\left(\frac{4}{3}\right)^2 + (6-4)^2}\right)$$

$$= 239 \text{ cm}^2$$

2. A bucket is in the form of a frustum. The diameters of the two ends are 14 cm and 21 cm. If the slant height of the bucket is 30 cm, calculate its surface area when it is (a) open   (b) closed

Solution

(a) $S = \pi(r^2 + (R+r)l)$

$$= \frac{22}{7}(7^2 + (10.5 + 7)30)$$

$$= 1804 \text{ cm}^2$$

(b) $S = \pi(R^2 + r^2 + (R+r)l)$

$$= \frac{22}{7}(10.5^2 + 7^2 + (10.5 + 7)30)$$

$$= 2150.5 \text{ cm}^2$$

7.6   Volume and Surface Area of Hemisphere

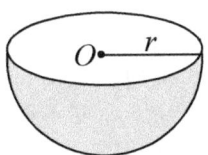

Since a hemisphere is half of a sphere, its volume V and curved surface area S are half that of the sphere. Hence,

$$V = \frac{2}{3}\pi r^3 \text{ and } S = 2\pi r^2$$

In addition, a hemisphere has a circular part so the total surface area of a solid hemisphere is

$$S = \pi r^2 + 2\pi r^2 = 3\pi r^2$$

Example

A hemispherical bowl with a flat lid has a radius of 7 cm. Calculate the
  (a) Volume of the bowl
  (b) The surface area of the bowl when
       (i) Open.   (ii) Closed

Solution

Volume of bowl $V = \dfrac{2}{3}\pi r^3$

$$\Rightarrow V = \dfrac{2}{3}\left(\dfrac{22}{7}\right)(7)^3 = 718.7 \text{ cm}^3$$

(b) (i) $S_0 = 2\pi r^2 = 2\left(\dfrac{22}{7}\right)(7)^2 = 308 \text{ cm}^2$

  (ii) $S = 3\pi r^2 = 3\left(\dfrac{22}{7}\right)(7)^2 = 462 \text{ cm}^2$

 Exercise 7:3

In this exercise, where necessary take $\pi = \dfrac{22}{7}$.

1. A bucket in the shape of a frustum has a base diameter of 4 cm and a brim diameter of 10 cm. Given that its height is 3 cm, calculate
    (a) The volume of the bucket.
    (b) The external surface area of the bucket.
2. A frustum cut out from a cone has diameters 6 cm and 18 cm and height 15 cm. Calculate
    (a) The volume of the frustum.
    (b) The external surface area of the frustum.
3. The figure below shows a frustum of a cone. The height of the frustum is 10 cm. If the diameters of the two ends are 8 cm and 6 cm respectively, calculate
    (a) The volume of the frustum,
    (b) The surface area of the frustum, leaving your answer in terms of π.

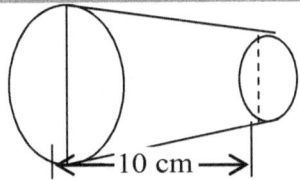

4. The figure below shows a sharpened round pencil with a hemispherical eraser of radius 0.5 cm. Calculate
   (a) The total volume of material contained in the pencil.
   (b) The total surface area of the pencil.

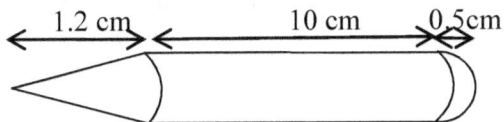

5. A cylindrical cake of radius 3 cm and height 4 cm is standing on a table. A slice of the cake is removed by cutting vertically through the radii OA and OB, where $\angle AOB = 30°$ as shown in the figure below. Calculate, leaving your answers in terms of π.
   (a) The volume of the remaining cake.   (b) Its total surface area.

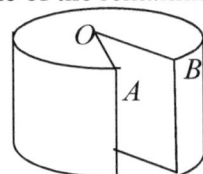

6. A cylindrical jar with internal diameter 15 cm contains a metal sphere of radius 6.3 cm. With the base of the cylinder horizontal, we pour water into it until it covers the sphere as in figure (a) below.
   (i) Calculate the fall in the water level after removing the sphere.
   (ii) Calculate the length of an edge of a metal cube immersed into the cylinder so that the water rises to its original level.

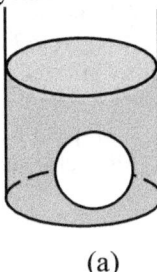   

(a)   (b)

7. Figure (b) above shows a cone of radius 4 cm attached to a hemisphere with equal radius. Given that the total height of the object is 10 cm, find
   (a) The volume of the object.   (b) The total surface area of the object.
8. Figure (a) below shows one corner of a solid cube of side 8 cm cut through the midpoints of three adjacent sides. Calculate the volume of the

remaining piece.

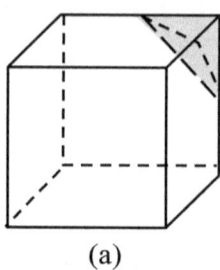
(a)

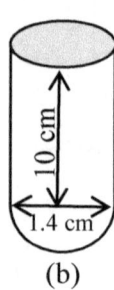
(b)

9. Figure (b) above shows a diagram of a test tube. Calculate
   (a) The volume of acid required to fill it right to the brim.
   (b) The total surface area of the test tube in contact with the acid when the test tube is full to the brim. (Assume that the thickness of the tube is negligible)
10. The external radius of a ball is 14 cm and the internal radius is 12.6 cm. Calculate to the nearest tenth of a centimeter,
   (a) The volume of air the ball can hold if fully inflated.
   (b) The volume of material that makes the ball.
   (c) The surface area of the ball.

Multiple Choice Exercise 7

In this exercises, where necessary, take $\pi = \frac{22}{7}$.

1. A cone made from a sector of a circle of radius 14 cm and angle 90° has a curved surface area of:
   [A] 22 cm²   [B] 88 cm²   [C] 77 cm²   [D] 154 cm²
2. The volume of a cone of radius 3.5 cm and vertical height 12 cm is:
   [A] 15.5 cm³   [B] 21.0 cm³   [C] 154.0 cm³   [D] 42.0 cm³
3. A sector cut off from a circle of radius 8.2 cm forms a cone of radius 3.5 cm. The curved surface area of the cone is:
   [A] 12.83 cm²   [B] 22.0 cm²   [C] 67.2 cm²   [D] 90.2 cm²
4. The angle of a sector of a circle of radius 8 cm is 240°. This sector can form a cone whose base radius is:
   [A] $\frac{16}{3}$ cm   [B] $\frac{15}{3}$ cm   [C] $\frac{16}{5}$ cm   [D] $\frac{8}{3}$ cm
5. The volume of a cone of height 9 cm is 1848 cm³. Its radius is:
   [A] 7 cm   [B] 14 cm   [C] 28 cm   [D] 98 cm
6. The total surface area of a cone whose height is 12 cm and whose base radius is 5 cm is:

[A] $240\frac{2}{7}$ cm² [B] $235\frac{5}{7}$ cm² [C] $282\frac{6}{7}$ cm² [D] $251\frac{3}{7}$ cm²

7. A cone is 14 cm deep and the base radius is $4\frac{1}{2}$ cm. The volume of water, which is exactly half the volume of the cone, is:
   [A] 49.5 cm³   [B] 99 cm³   [C] 148.5 cm³   [D] 297 cm³

8. The total surface area of a solid circular cone with base radius 3 cm and slant height 4 cm is:
   [A] 66 cm²   [B] $\frac{753}{7}$ cm²   [C] $\frac{782}{7}$ cm²   [D] 88 cm²

9. The base radius of a cone made from a sector of a circle of radius 21 cm and angle 210° is:
   [A] $3\frac{1}{2}$ cm   [B] 7 cm   [C] $10\frac{1}{2}$ cm   [D] $12\frac{1}{4}$ cm

10. The total surface area of a solid cone of slant height 15 cm and base radius 8 cm in terms of π is:
    [A] 120π cm²   [B] 184π cm²   [C] 200π cm²   [D] 320π cm²

11. The curved surface area of a cone of radius 3 cm and slant height 7 cm is:
    [A] 22 cm²   [B] 44 cm²   [C] 66 cm²   [D] 132 cm²

12. The height of a right circular cone is 4 cm. The radius of the base is 3 cm. Its curved surface area is:
    [A] 9π cm²   [B] 15π cm²   [C] 16π cm²   [D] 20π cm²

13. The base diameter of a cone is 14 cm, and its volume is 462 cm³. Its height is:
    [A] 3.5 cm   [B] 5 cm   [C] 7 cm   [D] 9 cm

14. The total surface area of a solid right circular cone of base radius r cm and height r cm is:
    [A] $2\pi r^2$ cm²   [B] $2\pi r^2$ cm²   [C] $\frac{7}{3}\pi r^2$ cm²   [D] $\frac{4}{3}\pi r^2$ cm²

15. The surface area of a sphere of radius 7 cm is:
    [A] 86 cm²   [B] 154 cm²   [C] 616 cm²   [D] 143 cm²

16. Two solid spheres have volumes 250 cm³ and 128 cm³ respectively. The ratio of their radii is certainly:
    [A] 5:4   [B] 25:16   [C] 2:1   [D] 4:3

17. A hollow sphere has a volume of k cm³ and a surface area of k cm². The diameter of the sphere is:
    [A] 3 cm   [B] 12 cm   [C] 9 cm   [D] 6 cm

18. A sphere has a surface area of 4312 cm². The radius of the sphere in cm correct to one decimal place is:
    [A] 18.0   [B] 18.5   [C] 19.0   [D] 19.5

19. The cross-section of a prism is a right-angled triangle 3 cm by 4 cm by 5 cm. The height of the prism is 8 cm. Its volume is:
    [A] 48 cm³   [B] 60 cm³   [C] 96 cm³   [D] 120 cm³

36-37 The following shows a triangular prism of length 7 cm. The right-angled triangle PQR is a cross section of the prism $|PR| = 5$ cm and $|RQ| = 3$ cm.

20. Use the information to answer questions 36 to 37.

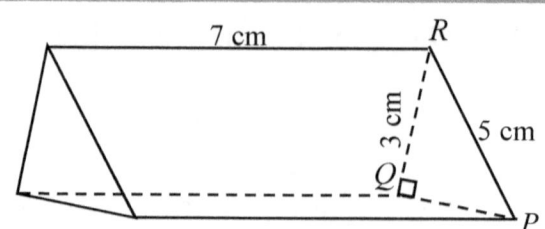

21. The area of the cross-section is:
    [A] 4 cm^2 [B] 6 cm^2 [C] 15 cm^2 [D] 20 cm^2
22. The volume of the prism is:
    [A] 28 cm^3 [B] 42 cm^3 [C] 70 cm^3 [D] 84 cm^3
23. The height of a pyramid on a square base is 15 cm. Given that the volume is 80 cm^3, the length of the side of the base in cm is:
    [A] 3.3   [B] 5.3   [C] 4.0   [D] 8.0
24. The height of a pyramid on a square base is 15 cm. If the volume is 80 cm^3, the area of the square base is:
    [A] 16 cm^2 [B] 9.6 cm^2 [C] 8 cm^2 [D] 25 cm^2
25. A right pyramid is on a square base of side 4 cm. The slanting side of the pyramid is $2\sqrt{3}$ cm. The volume of the pyramid is:
    [A] $\dfrac{10}{3}$ [B] $\dfrac{16}{3}$ [C] $\dfrac{32}{3}$ [D] $\dfrac{64}{3}$
26. A pyramid on a square base of side 10 cm has a height of 15 cm, its volume must be:
    [A] 150 cm^3 [B] 500 cm^3 [C] 1500 cm^3 [D] 5000 cm^3
27. The base of a pyramid is a 12 cm by 12 cm. If its height is 20 cm, the volume of the pyramid in cm^3 is:
    [A] 960   [B] 80   [C] 1440   [D] 1600
28. The figure below shows a cone with the dimensions of its frustum indicated. The height of the cone is:
    [A] 12 cm   [B] 15 cm   [C] 18 cm   [D] 24 cm

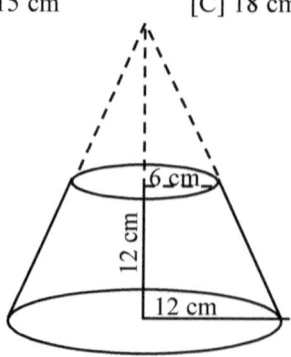

Module 13
Statistics and Probability

Family of Situations

At the end of module 13, the student is expected to have acquired more competencies within the families of situations *'Organization of information and estimation of quantities in the consumption of goods and services'*.

Categories of Action

The categories of action for module 13 include:
1. Organization, presentation and exploitation of information
2. Interpretation of results.
3. Taking chances.

Credit

The module is expected to be covered within 3 weeks teaching 4 hours per week (or within 10 to 12 hours).

Topic 8

STATISTICS

Objectives
At the end of this topic, the learner should be able to:

1. Draw up frequency distribution tables.
2. Represent data using a bar chart, a pie chart or a histogram.
3. Read and interpret data from charts.
4. Find class width, mid class value or centre.
5. Find the measures of central tendencies for given data.
6. Construct histograms for group data.

Module 13, Topic 8: Statistics

8.1 Frequency-Distribution Tables

In books 1 and 2 we saw that a **frequency distribution table** or simply a **frequency distribution** is tabular summary of statistical information and we call the process of constructing frequency distribution tables' **tabulation**. **Frequency** usually denoted by f is the number of times a particular statistical entity, x, occurs.

8.2 Representation of Data-Statistical Graphs

The various ways of representing data include pictograms, bar charts, pie charts, histograms etc. The names 'chart', 'graph', or 'diagram' are synonymous in this context. In this topic we shall use these words interchangeably.

(i) Pie Charts or Circular Diagrams

In this type of graphs, we divide a circle into sectors. The size of each sector is proportional to the frequency of the statistic it represents. Therefore necessary we should always draw pie charts with a pair of compass and measure the angles with a protractor. To draw a pie chart we first calculate the angles as follows.

Example

A trader bought 200 pineapples, 400 Bananas, 500 watermelons and 900 Oranges. Represent this information on a pie chart.

Solution

Fruit	No.	Calculation	Angle
Pineapple(P)	200	$\frac{200}{2000} \times 360$	$36°$
Bananas (B)	400	$\frac{400}{2000} \times 360$	$72°$
Watermelons (W)	500	$\frac{500}{2000} \times 360$	$90°$
Oranges (O)	900	$\frac{900}{2000} \times 360$	$162°$
TOTAL	2000		$360°$

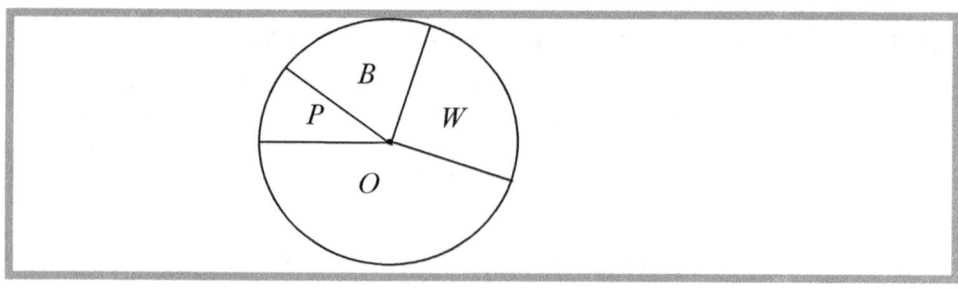

(ii) Bar Charts

In this case, bars of equal width represent information. Only the length or height of the bars has any significance. Therefore, we can draw the width of the bars any convenient size. The bars may be together or separated but must not overlap. Bar charts are of many types. We normally draw bar charts horizontally or vertically.

(a) *Chronological Bar Chart*

These types of bar chart, which may be vertical or horizontal usually, represent the variation of some statistic over a period.

Example

The following shows the number of people who owned televisions in a certain town from 1985 to 1993. Draw a chronological bar chart to represent this data.

Year	Number of People
1985	600
1986	800
1987	900
1988	1100
1989	1300
1990	1400
1991	1200
1992	1500
1993	1600

Solution

Number of TV Owners from 1985 to 1993

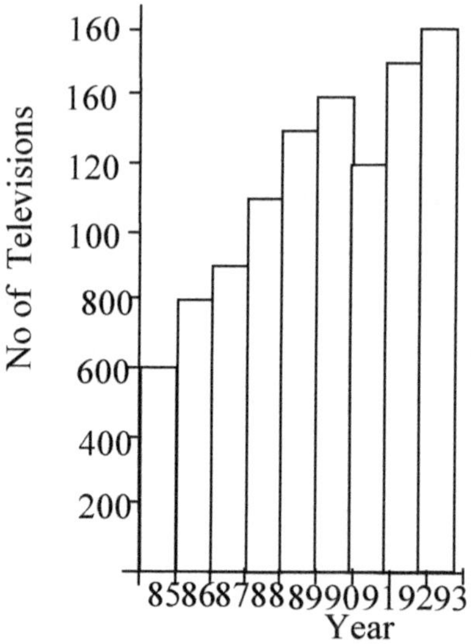

(b) *Simple Bar Charts*
The bars in this type of bar chart separate from each other.

Example

Use the table above to draw a simple bar chart.

Solution

OR

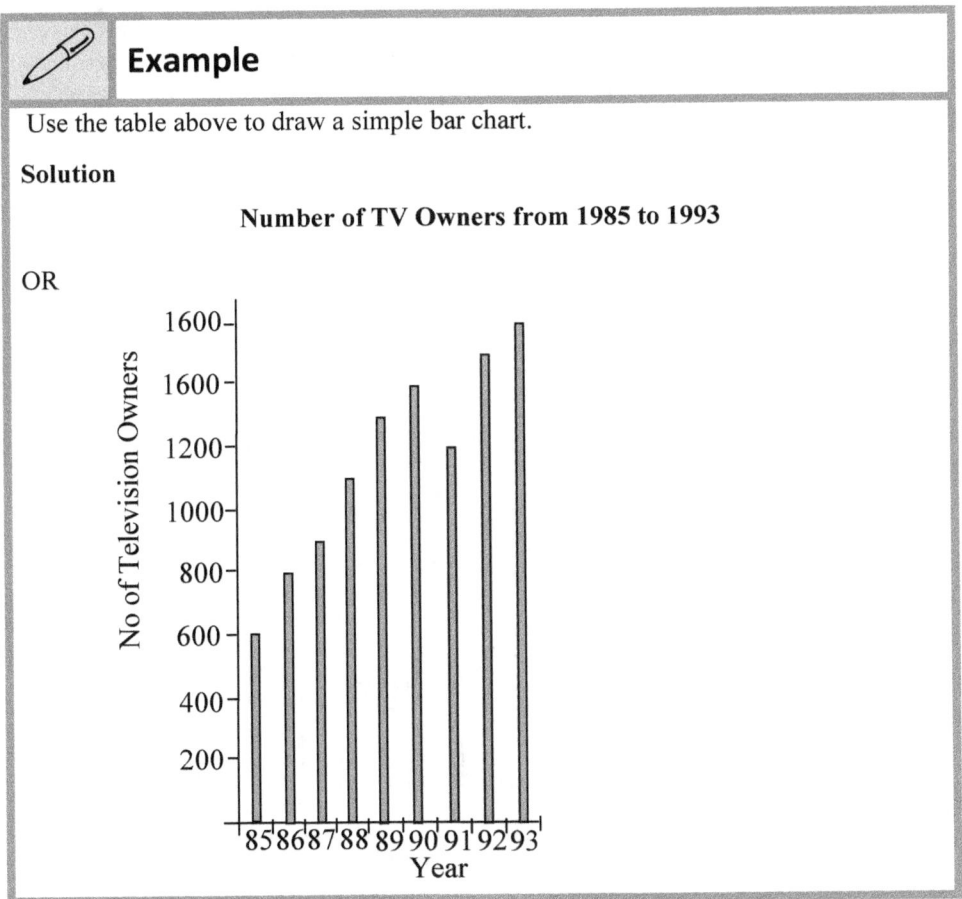

(iii) Histograms

Histograms are similar to chronological bar charts. The only difference is that while the frequency is proportional to the length of the bars in the case of the bar chart, the frequency is proportional to the areas of the rectangles in the case of the histogram. Therefore, for the histogram, both the length and the width of the rectangle are important. Though the width of the rectangle, may be of different sizes it is often more convenient to make them the same size. In this way, histograms are therefore very similar to chronological bar charts

Example

The table below shows the number of form 2 students in a certain school in the year 2002 who had the required textbooks for the subjects Mathematics (M), English (E), French (F), History (H), Geography (G), Chemistry (C), Physics (P), Biology (B) and Literature (L).

Draw a histogram to represent this information.

Textbook	M	E	F	H	G	C	P	B	L
No. of students	14	15	11	3	9	5	7	1	13

Solution

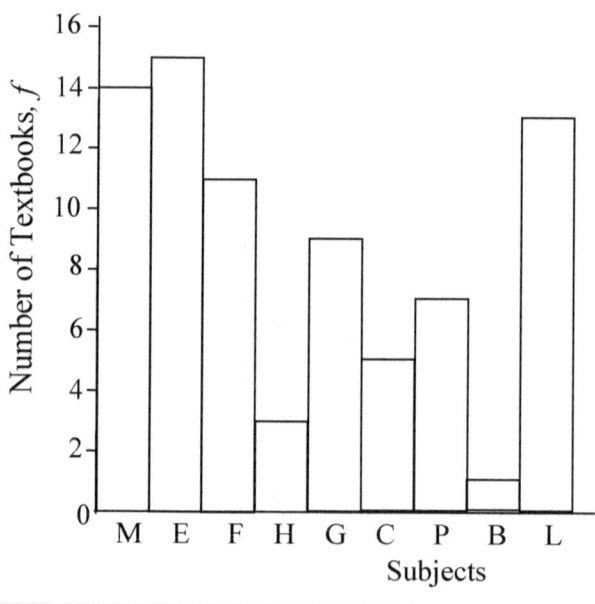

Module 13, Topic 8: Statistics

 Exercise 8:1

1. The form three students of a certain school participated in team sports as shown in the table below.

Team sport	Number of students
Handball	45
Basketball	60
Football	75

   Represent this information on a pie chart and state the angle for basketball.

2. The following is a pie chart (not drawn to scale) showing how a student spent his pocket money amounting to 27,000 FCFA. Given that he spent twice as much on books as he did on taxi, calculate:
   (a) How much he spent on books.
   (b) How much he spent on others.

   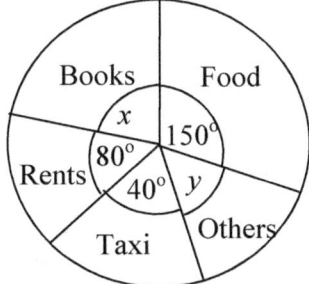

3. The table below shows a survey carried out on a group of students to find out what they ate for launch on a certain day.

Achu	15
Rice	9
Garri	4
Bread	2

   Draw a histogram to display this data

4. The pie chart below shows the number of votes for candidates A, B and C in an election. Calculate the percentage of the votes to the nearest tenth in favour of candidate B.

   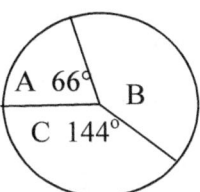

5. Draw a histogram for the distribution in the table below.

Score, x	1	2	3	4	5	6
Frequency, f	4	6	7	3	3	1

7. The livestock of a certain farm consist of 28 cows, 300 sheep, 74 pigs, 306 poultry, 9 dogs and 3 cats. If we are required to record this information on a pie

chart, calculate the angle in degrees, at the centre of the sector representing the cows.

8. Five boys A, B, C, D and E are of heights 160, 144, 120, 96 and 80 centimetres respectively. Represent this information on a bar chart.

9. The figure below shows a pie chart indicating the favourite colours of a group of 108 girls.

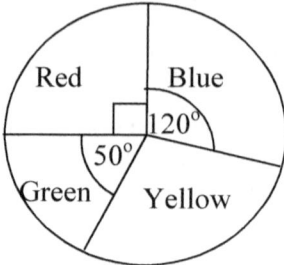

(a) Find the angle of the sector for girls who like yellow.
(b) Find the number of girls who like green.

10. A student used the following table to draw a pie chart. Find the values of w, x, y and z.

Item type	A	B	C	D
Frequency	48	104	x	y
Sector angle (°)	72	w	108	z

11. The following shows a statistical table of a variable x with frequency f. We need to represent this data on a pie chart. Calculate in degrees, the angle of the sector representing $x = 3$.

x	1	2	3	4	5	6
f	4	6	7	3	3	1

8.3 Measures of Central Tendencies

Mode, mean and **median**, which are very commonly used, are examples of averages, otherwise called measures of central tendencies. We call them measures of central tendencies because their values are representative or typical of any given data, and tend to lie centrally when we rank or arrange the data in order of magnitude (from highest to smallest or smallest to highest). Each of these measures has its advantages and disadvantages depending on the data and it intended purpose. For instance, the mean has the disadvantage that extreme values strongly affect it. Extreme values do not affect the median. On the other hand, the mode may or may not exist and turns to be very subjective. At times, there are many modes.

Mode

The mode of any given data is the variable or statistic that occurs most frequently.

Module 13, Topic 8: Statistics

 Example

1. Find the mode of the following data. 1,2,4,6,2,7,7,2,2,7.

   Solution

x	1	2	6	7
f	1	4	1	3

   The frequency of 2 is 4, which is the highest frequency, so 2 is the mode.

2. Find the mode of the following data.
   (i) 99, 100, 101, 102 and 101.   (ii) 99, 100, 101, 100, 102 and 101.

   Solution
   (i) Mode = 101   (ii) Mode = 100 and 101

Notice in the example above that there are 2 modes, 100 and 101. We call such a distribution a **bimodal** distribution. If there are three modes, we call the distribution a **trimodal** distribution and generally, if there are many modes we call the distribution a **multimodal** distribution.

8.4 Arithmetic Mean (Average or Mean)

We denote the mean by $\bar{x}$ and obtain it by summing all the data and dividing by the frequency. Thus

$$\bar{x} = \frac{\text{sum of data}}{\text{total frequency}}$$

$$\bar{x} = \frac{\sum x}{\sum f}$$

Where $\sum x$ and $\sum f$, read 'summation x' and 'summation f', respectively meaning 'sum of data' and 'sum of the frequencies' respectively. $\sum$ is called the sigma notation and $\sum x$ and $\sum f$, are read 'sigma x' and 'sigma f' respectively.

 Example

1. Find the mean of 11, 9, 15, 12 and 13

Solution

$$\bar{x} = \frac{\sum x}{\sum f} = \frac{11+9+15+12+13}{5} = 12$$

If some of the data repeat themselves, we take advantage of multiplication as repeated addition, to write the formula as

$$\bar{x} = \frac{\sum xf}{\sum f}$$

$\sum fx$, is read 'sigma fx', or 'summation fx', where fx means the product of each statistic and its frequency.

2. The following shows the marks obtained by 30 students during a test. Calculate the average mark.

55	60	65	40	60	60
65	50	40	60	50	60
60	50	60	30	40	60
60	50	60	50	60	50
60	50	60	60	50	60

Solution
To ease the work we draw a frequency distribution table.

Mark, x	Frequency, f	fx
30	1	30
40	3	120
50	8	400
55	1	55
60	15	900
65	2	130
	$\sum f = 30$	$\sum fx = 1635$

$$\bar{x} = \frac{\sum fx}{\sum f} = \frac{1635}{30} = 54.5$$

3. Find the mean of the following data.
13, 13, 13, 13, 13, 13, 14, 14, 15, 15, 15, 16, 16, 16, 16, 16, 16, 16, 16, 16

Solution

x	f	fx
13	6	78
14	2	28
15	3	45
16	9	144
	$\sum f = 20$	$\sum fx = 295$

$$\bar{x} = \frac{\sum fx}{\sum f} = \frac{295}{20} = 14.75$$

8.5 Median

To obtain the median, first rank the data. In other words, arrange the data in order of magnitude. For an odd number of numbers, the median is the middle number and for an even number of numbers, the median is the average of the two middle numbers.

Example

1. Find is the median of 12,2,7,13,6.

   Solution

   Ranking: 2, 6, 7, 12, 13
   ∴Median = 7

2. Find the median of 2, 7, 6, 13, 12, and 8

   Solution

   Ranking: 2, 6, 7, 8, 12, 13

   $$\therefore \text{median} = \frac{7+8}{2} = 7.5$$

Exercise 8:2

1. Find the number that must be removed from the eight numbers 4, 11, 13, 8, 4, 5, 8 and 2, so that the mean of the remaining seven numbers is 6.
2. The mean of five numbers is 4. When we add a sixth number, the mean of the six numbers is $3\frac{1}{2}$. Find the sixth number.
3. Given that the mean of 3, 4 and m is 6, find the mean of 2, m and 14.
4. The table below represents the weights in kg of 11 students.

Weight, kg	45	53	54	49
No. of students	2	3	4	2

   (a) State the modal weight of the students.
   (b) Find the mean weight of the students.
   (c) Find the median of the distribution.
5. Use the frequency distribution in the following table to calculate:
   (a) the mean (b) the modal score (c) Find the median of the distribution.

Score (x)	1	2	3	4	5	6
Frequency (f)	4	6	7	3	3	1

6. The table below shows the marks obtained by pupils in a mathematics test.

Marks (x)	0	3	5	6	8	9	10
No. of pupils (f)	2	4	6	2	4	1	1

   (a) State the mode of the distribution.
   (b) Calculate, to 1 decimal place, the mean of the distribution.
   (c) Find the median of the distribution.
7. Consider the frequency distribution in the following table.

Score x	3	5	7	9	11
Frequency f	4	6	10	5	5

   (a) State the mode of the distribution.
   (b) Calculate, to 1 decimal place, the mean of the distribution.
   (c) Find the median of the distribution.
8. The table below shows the number of coins of six denominations in a bag. Find:
   (a) the average value of the coins in the bag.
   (b) the mode of the coins in the bag.
   (c) the median of the distribution.

Value of coin FRS	5	10	25	50	100	500
Number	4	10	6	8	15	7

9. The weights of 8 teachers in a certain primary school were measured in kg as follows: 74, 64, 68, 76, 80, 72, 68 and 60 respectively. Find
   (a) the median. (b) the mode of the data. (c) their mean weight.
10. The frequency distribution in table below shows the scores in a mathematics test in a certain class.

Score (x)	2	3	4	7	8	9
Frequency (f)	1	4	6	8	9	2

(a) Find how many students wrote the test.
(b) Find the mode of this distribution.
(c) Find the mean mark for the test to 1 d.p.

11. The numbers of absences from a mathematics class registered within the first 20 lessons in the first term are 2, 3, 1, 0, 0, 4, 3, 2, 2, 2, 1, 4, 5, 5, 0, 0, 1, 1, 2, and 2. Find the
    (a) mode   (b) median   (c) mean number of absences.

12. 10 packets of different sizes contain sweets as shown in the table below.

Number of sweets	5	12	6	15
Number of packets	4	2	3	1

(a) State the mode of the number of packets.
(b) Find the median of the number of packets.
(c) Calculate the mean number of sweets per packet.

8.6 Choicest Measure of Central Tendency

In choosing which measure of central tendency to use, we must consider two things. These are:

1. The nature of the distribution,
2. The purpose we intend to use the measure of central tendency.

Mean

The mean has the advantage that:
(a) We can express it using a simple formula.
(b) It takes account of all the values involved in the distribution.
   The disadvantage of the mean is that it is highly affected by extreme values. Therefore, it is not a good measure of central tendency if the distribution involves one or more extreme values in one direction.
   For instance in the data 3, 5, 8, 36, the mean is 13. This will be a very misleading measure of central tendency.

Median

The median has the advantages that:
(a) It is very easy to calculate.
(b) It is the middle of any distribution and extreme values do not affect it.
   The median has the disadvantage that too many values in one direction greatly affect it.
   In the data 1, 3, 4, 6, 17, 18, and 19, the median is 6. This value is too low because most of the values are small. Therefore, the median is not a good representation of the distribution in this case.

Mode

The mode has the advantage that:
(a) It is far easier to determine.
(b) Extreme values do not affect it.

The disadvantage of the mode is that, it is meaningless when there are several values having the highest frequency of occurrence. For instance
$$2, 5, 3, 2, 2, 7, 5, 8, 5, 7, 3, 9, 7, 3$$
Ranking and tabulating the data

x	2	3	5	7	8	9
f	3	3	3	3	1	1

The values 2, 3, 5, and 7 by definition are the modes, which has no significance.

Example

1. What is the most appropriate measure of central tendency for the following distribution?   3, 4, 4, 4, 4, 4, 6, 7, 9, 13

Solution
The median is 4, the mode is 4, and the mean is 5.8.
The mean is the best because it is almost central and therefore the best representation of the data.

2. The hourly wages of five employees in an office are 252 FRS, 396 FRS, 328 FRS, 920 FRS and 325 FRS. Find
   (a) The median hourly wage.   (b) The mean hourly wage.
   (c) Comment on your result.

Solution
(a) Arranging in ascending order, the wages are 252 FRS, 328 FRS, 375 FRS, 396 FRS and 920 FRS. This gives the median of 375 FRS.
(b) $\bar{x} = \frac{252+396+328+920+325}{5} = 444.2$
(c) The extreme value 920 FRS does not affect the median but affects the mean. In this case, the median is a better measure of the average hourly wage than the mean.

Exercise 8:3

1. Which measure of central tendency do you think the manager of New Life Supermarket will be most interested in? Give reasons for your answer.
2. In a certain week, a bus driver brought to his Patron the following balances: 20,000 FCFA, 20,000 FCFA, 22,000 FCFA, 24,000 FCFA, 24,000 FCFA, 36,000 FCFA, and 38,000 FCFA respectively on each day of the week. What measure of central tendency would you use to compare the balances, and why?
3. The data in the table below shows the number of students of LS1, who passed in the following subjects: Mathematics (M), Physics (P), Chemistry (C), Biology (B) and Further Mathematics (F).

Subject	M	P	C	B	F
No. of students	8	7	2	8	1

Which measure of central tendency would be the most appropriate for the analysis of this data? Give reasons for your answer.

4. The following table is a survey of the number of pigs owned by a group of farmers. One of Nga's two pigs just had a litter of 7 and 8 piglets respectively, so he has 16 pigs.

Number of piglets	0	1	2	4	16
Number of students	10	10	6	3	1

(a) Explain why the average should not be used as a measure of central tendency in this case.
(b) State the most appropriate measure of central tendency required in this case.

8.7 Grouped Data

To analyse very large masses of raw data, it is often necessary to distribute the data into **classes, class intervals** or **groups,** as shown in the table below, which is frequency distribution of the scores of 50 students in a test. We call this grouped data.

Mark, x	30-39	40-49	50-59
Frequency, f	10	14	26

Though data grouped in this way is easier to analyse, this method has a disadvantage in that the grouping destroys most of the original details of the information.

Class Limits and Class Boundaries

Consider the class 30-39. We call the smaller number, 30 the **lower class limit**, and the larger number 39, the **upper class limit**. Thus, 40 and 50 are the lower class limits, while 49 and 59, are the upper class limits for the classes 40-49 and 50-59 respectively.
If the processor rounded the data to the nearest whole number, the true class limits (called **class boundaries**) will actually be 29.5, 39.5, 49.5 and 59.5. For the class 30-39, the lower and upper class boundaries will be 29.5 and 39.5 respectively. For the class 40-49, the lower and upper class boundaries will be 39.5 and 49.5 respectively.

Class Size

This is the difference between the upper class boundary and the lower class boundary. Other names for class size are **class width, class length** or class interval denoted by c.

Mid-Interval Value

Other names for the mid-interval value are class mark or mid-point. The mid-interval value is the arithmetic mean of the upper and lower class limits. Thus for the class 40-49,

$$\text{Class mark} = \frac{40+49}{2} = 44.5$$

The class mark is the mark, which is representative of the given class.
Therefore, for clarity and mathematical analyses, we assume that all observations within a given class coincide with the class mark.

8.8 Histograms for Grouped Data

When drawing histograms for grouped data, it is preferable to use the class boundaries and the class mark than the class limits.

(a) *Histograms with equal class widths*

These types of histograms are the most commonly used. In this case, the frequency is proportional to the height (length) of the rectangles.

Module 13, Topic 8: **Statistics**

Example

The table below shows the marks obtained by 80 students in an examination.

Marks, x	Frequency, f
1-10	3
11-20	5
21-30	5
31-40	9
41-50	11
51-60	15
61-70	14
71-80	8
81-90	6
91-100	4

Draw a histogram to represent this data.

Solution
Marks Obtained by 80 Students in an Examination

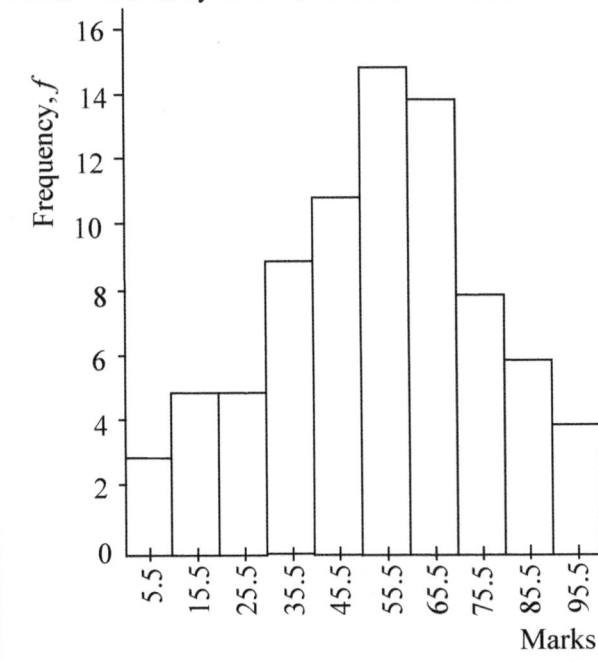

Finding the Mode from a Histogram

The Modal class is the class with the highest frequency represented by the tallest bar in the histogram. We can obtain the mode from a histogram as in the example below.

Marks Obtained by 80 Students in an Examination

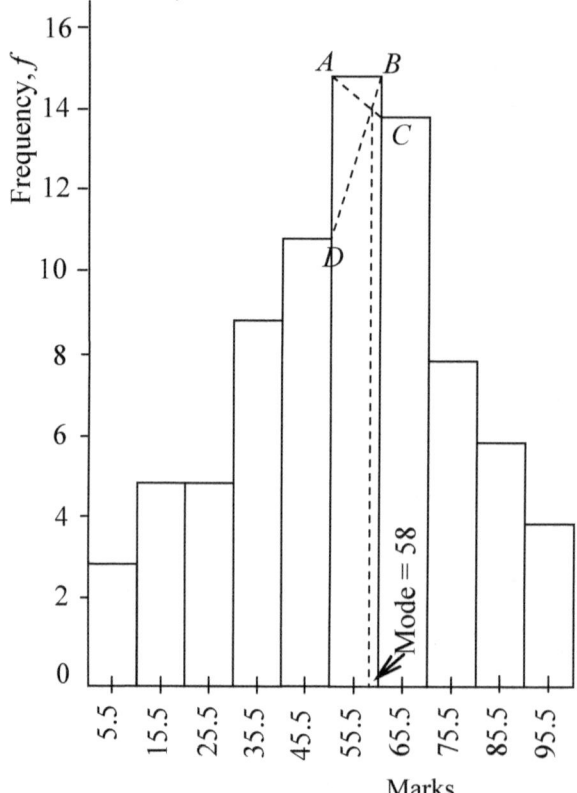

 Example

Use the histogram in the example above to obtain the mode.

Solution

To obtain the mode we read the mark corresponding to the point of intersection of the dotted lines *AC* and *BD* as shown in the figure above. From the figure above, the mode is 58.5.

8.9 Finding the Mode by Calculation

If the class intervals are of equal sizes, we can obtain the mode using the formula

$$\text{mode} = L_1 + \left(\frac{\Delta_1}{\Delta_1 + \Delta_2}\right)c, \text{ where}$$

L_1 = lower class boundary of the modal class
C = class width
Δ_1 = modal class frequency − next lower class frequency
Δ_2 = next upper class frequency − modal class frequency

 Example

By calculation, find the mode of the data in the example above.

Solution

$$\text{mode} = L_1 + \left(\frac{\Delta_1}{\Delta_1 + \Delta_2}\right)c = 50.5 + \left(\frac{4}{4+1}\right)10 = 58.5$$

8.10 Frequency Distribution Curve (Frequency Polygon)

After drawing a histogram we can draw, another type of graph called a **frequency distribution curve** or **frequency polygon** (shown in the next example) by joining the tips of the rectangles of the histogram.

 Example

1. Use the histogram in above to draw a frequency distribution curve.

Solution

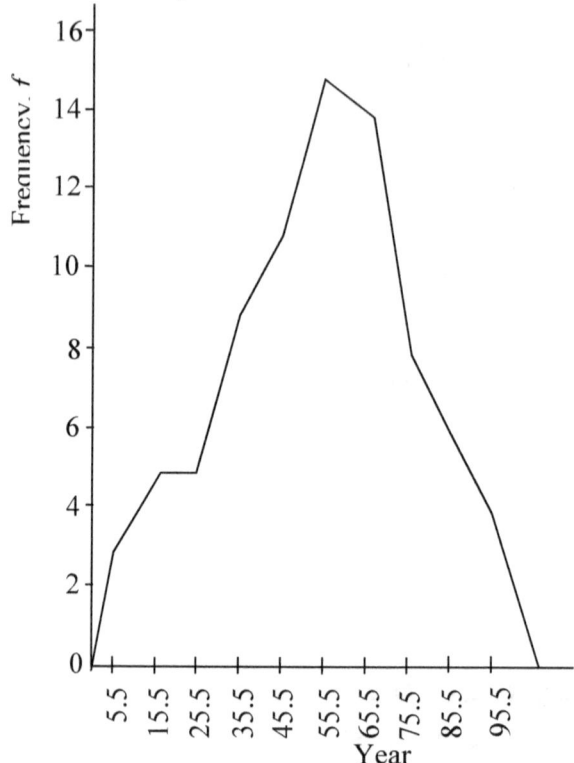

Marks Obtained by 80 Students in an Examination

2. Thirty-six students obtained the following scores out of 100 in a test.

```
25  49  76  12  51  56
81  50  45  92  58  67
55  52  43  31  48  84
66  56  44  39  45  22
56  74  98  67  34  41
34  68  69  70  85  51
```

(a) Starting with 0-9, arrange the marks in a grouped frequency table with class intervals of size 10.
(b) State the modal class.
(c) Draw a histogram to represent this data, hence obtain the mode of the distribution.
(d) Draw a frequency polygon of the distribution.

Solution

(a)

Marks, x	Class mark	Frequency, f
0-9	4.5	0
10-19	14.5	1
20-29	24.5	2
30-39	34.5	4
40-49	44.5	7
50-59	54.5	9
60-69	64.5	5
70-79	74.5	3
80-89	84.5	3
90-99	94.5	2

(b) The modal class is 50-59

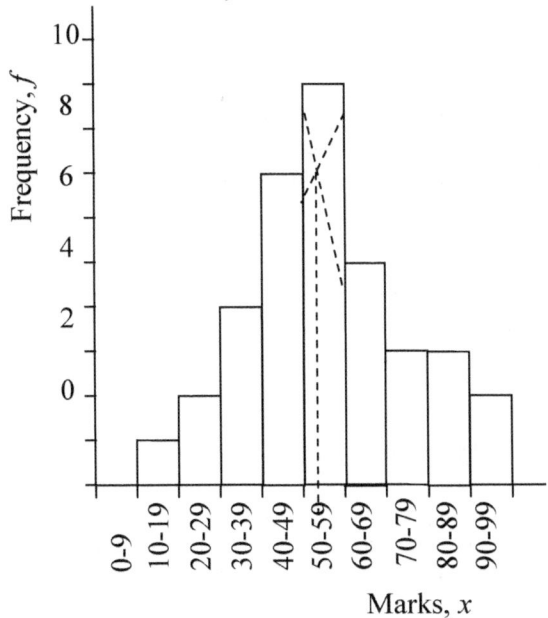

Marks Obtained by 36 Students in a test

(c) From the histogram, the mode is 53.5.
(d) The frequency polygon is as shown below.

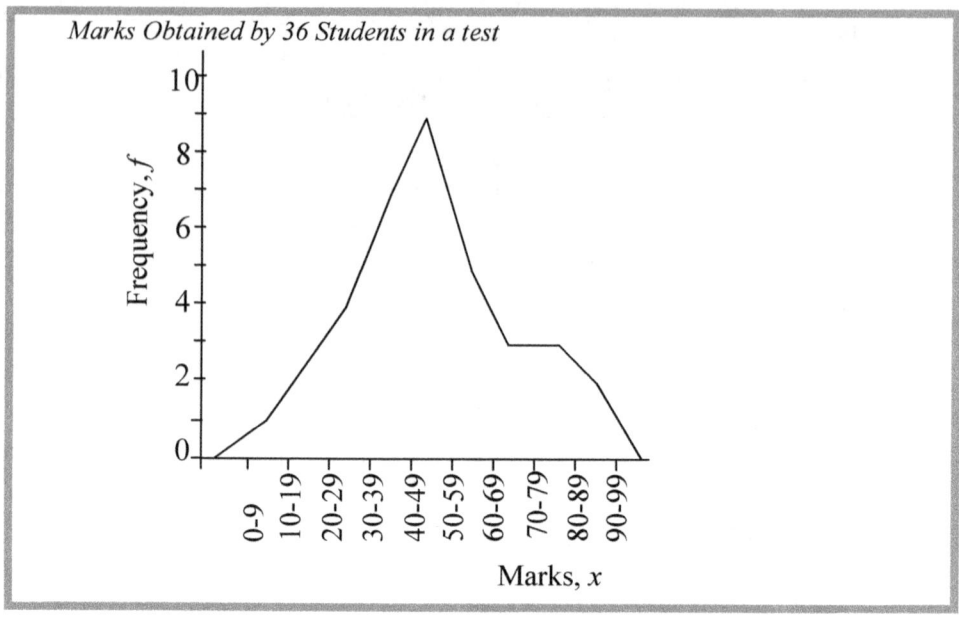

Marks Obtained by 36 Students in a test

(b) *Histograms with unequal class widths*

Histograms with unequal class widths are less common than histograms with equal class widths. However, it is important that we should take note of them. On such histograms,

Frequency = class width × standard frequency,

Or

Standard frequency, S.F. = $\dfrac{\text{frequency of the class}}{\text{class width}}$

Another name for standard frequency is relative frequency or frequency density and we can represent by the height of each rectangle. Therefore, the frequency is proportional to the area of the rectangle for each class and not to the height of each bar.

Example

The following table shows the wage distribution amongst three groups of employees in a large company.
(a) Draw a histogram for the distribution.
(b) Draw the frequency polygon for the distribution.

Wage in thousand CFA	Frequency
0-9	8
10-19	16
20-39	10

Solution

x	f	S.F.
0-9	8	0.8
10-19	16	1.6
20-39	10	0.5

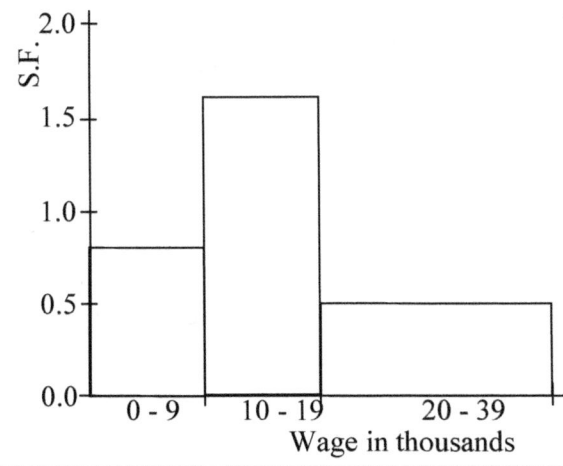

163

Exercise 8:4

1. A boy measured to the nearest metre, how far he could throw tennis on 20 successive trials, and obtained the following results:
   66 69 70 68 71 68 69 70 67 68
   68 68 67 66 69 68 69 70 68 67
   (a) Draw a frequency distribution table for the data.
   (b) State the mode of the distribution.
   (c) Find the median of the distribution
   (d) An approximate formula for determining the mean is
   Mean − Mode = 3(Mean − Median)
   Use this formula to calculate the mean of the distribution.
   (e) Calculate the exact mean of the distribution, to one decimal place.

2. $\frac{1}{9}$ of the candidates obtained grade D and an equal number of the candidates were awarded grades E and F.
   (a) Copy and complete the following table that is used to draw up a pie chart.

GRADE	A	B	C	D	E	F
Angle of Sector					70	70

   (b) Draw the pie chart accurately on a circle of radius 5 cm, labeling the sectors.
   (c) Given that 6 candidates obtained grade A,
      (i) Find the number of candidates that took the test.
      (ii) Find the number that obtained grade D.
   (d) Given that grades A, B, and C, are pass grades, state the ratio 'pass to fail' in its simplest form.
   (e) The figure below shows the distribution of the grades for boys. On a similar set of axes, construct a bar chart to show the distribution of the grades for girls.

   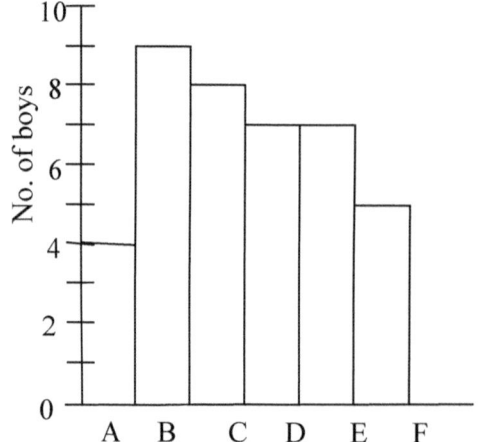

3. A mathematics teacher gave a test and promised to award prizes to the top 30

Module 13, Topic 8: Statistics

students of the class. The marks scored by the top 30 students were recorded as follows:
10, 8, 4, 4, 5, 7, 4, 6, 5, 4, 9, 7, 6, 4, 8, 8, 6, 7, 4, 6, 8, 5, 6, 4, 8, 7, 6, 4, 6, 4.
   (a) draw a frequency table to show the distribution of marks over the top 30 students in the class
   (b) state the mode of the distribution
   (c) Calculate the mean mark for the top 30 students.
4. The mean of two numbers in a set A is 21.8. The mean of three numbers in a set B is 28.8. Find:
   (a) the sum of the two numbers in set A,
   (b) the sum of the three numbers in set B,
   (c) the mean of the five numbers put together.
   (d) When we introduce a sixth number x into the set in (c) above, the mean of the six numbers is now 20, find x.

Multiple Choice Exercise 8

1. We call a survey of people and/or property:
   [A] a census   [B] data   [C] a population   [D] a sample
2. The tally marks |||| |||| |||| |||| |||| ||| represent the number:
   [A] 18   [B] 23   [C] 28   [D] 33
3. The correct tally representation of 17 students is:
   [A] |||| |||| |||| [B] |||| |||| ||||||| [C] ||||| |||| |||| || [D] |||| |||| |||| ||
4. The tally marks |||| |||| |||| ||| stand for:
   [A] 18   [B] 15   [C] 13   [D] 20
5. A school awarded 480 out of 720 candidates a D Grade. On a pie chart showing all the grades the angle at the centre for the D grade is:
   [A] 270°   [B] 240°   [C] 210°   [D] 180°
6. The measure of central tendency a shoe company will be most interested in is:
   [A] mean   [B] mode   [C] median   [D] mean and median
7. The average of 0, 1, 6, 7, 9 and 19 is:
   [A] 9   [B] 6   [C] 7   [D] 10
8. The average of 1, 2, 5, 7, and 15 is:
   [A] 6   [B] 30   [C] 7   [D] 15
9. A group of four people found that their heights were 1.38 m, 1.71 m, 1.23 m and 1.40 m. Their average height (in metres) is:
   [A] 1.145   [B] 1.18   [C] 1.39   [D] 1.43
10. The average wage bill in FCFA of 40 men who collectively earn 3,540,000 FCFA is:
    [A] 87,000   [B] 29,500   [C] 88,500   [D] 31,700
11. The mean of 9, 13, 16, 17, 19, 23, 24 correct to two decimal places is:
    [A] 23.00   [B] 17.29   [C] 16.50   [D] 16.33
12. The average of the first four prime numbers greater than 10 is:

[A] 20 [B] 19 [C] 17 [D] 15
13. The mean of 20 observations is 4. The observed largest value 23 is removed. The mean of the remaining observations is:
    [A] 4 [B] 3 [C] 2.85 [D] 2.60
14. The mean heights of the three groups of students consisting respectively of 20, 16 and 14 students are 1.67 m, 1.50 m and 1.40 m respectively. The mean height of all the students is:
    [A] 1.52 m [B] 1.53 m [C] 1.54 m [D] 1.55 m
15. The mean of 30 observations is 5. The observed largest value of 34 is deleted. The mean of the remaining observations is:
    [A] 4 [B] 3.8 [C] 3.4 [D] 5
16. The following table shows the scores of some students in a test. The average score is 3.5. The value of x is:

Scores	1	2	3	4	5	6
No. of students	1	4	5	6	x	2

    [A] 1 [B] 2 [C] 3 [D] 4
17. The mean of 9, x and 13 is 11. The value of x is:
    [A] 7 [B] 8 [C] 11 [D] 13
18. The value of x which qualifies 4 as the mean of the data 4, $3x$, 0 and 3 is:
    [A] 1 [B] 2 [C] 3 [D] 4
19. The mean of five observations is 15. Four of them are 11, 12, 19 and 20. The fifth is:
    [A] 10 [B] 25 [C] 20 [D] 13
20. A pie chart is drawn to represent the percentages 20%, 50%, 25% and 5%. The angle which represents 5% is:
    [A] 5° [B] 18° [C] 25° [D] 126°
21. Given the scores –3, 4, 0, 4,–2,–5, 1, 7,10,5 the median of the scores is:
    [A] 2.5 [B] 2 [C] 4 [D] 3.5
22. From Table 37:52, the mean number of male children per family is:
    [A] 5 [B] 4 [C] 3 [D] 2

No. of male children	0	1	2	3	4
No. of families		4	8	6	2	7

23. The mean of four numbers a, b, c and d is 6. The mean of 5 numbers a, b, c, d and e is 10. The value of e is:
    [A] 24 [B] 25 [C] 26 [D] 27
24. The average age of five boys is 11 years. A sixth boy whose age is 17 years is added, the mean age in years will now be:
    [A] 14 [B] 12 [C] 13 [D] 11
25. The median of 8, 10, 9, 6, 7, 10, 12, 8, 9, 8 is:
    [A] 7.5 [B] 8 [C] 8.5 [D] 8.7
26. The median of the set of scores 65, 75, 55, 48, 78 is:
    [A] 55 [B] 60 [C] 72 [D] 65
27. The median of the set of numbers 2.64, 2.50, 2.72, 2.91, 2.35 is:
    [A] 2.72 [B] 2.64 [C] 2.50 [D] 2.35

28. Given the set of numbers 12,15,13,14,12 and 12. The median is:
    [A] 12.5   [B] 12   [C] 13   [D] 13.5
29. The following table shows the age distribution of a group of children. Their median age is:
    [A] 4 years   [B] 7 years   [C] 8 years   [D] 9 years

Age (in years)	4	5	6	7	8	9	10
Frequency	2	1	2	4	3	6	2

30. The following table gives the marks scored by a group of students in a test. The median mark is:
    [A] 4   [B] 3   [C] 2   [D] 1

Mark	0	1	2	3	4	5
Frequency	1	2	7	5	4	3

31. The mode of the numbers 8, 10, 9, 9, 10, 8, 11, 8, 10, 9, 8 and 14 is:
    [A] 8   [B] 9   [C] 10   [D] 11
32. A group of students measured a certain angle (to the nearest degree) and obtained the following results.
    75° 76° 72° 73° 74° 79° 72°
    72° 77° 72° 71° 70° 78° 73°
    The mode of their measurements is:
    [A] 78°   [B] 74°   [C] 73°   [D] 72°
33. The measure, which is not a measure of dispersion, is:
    [A] Mode   [B] mean deviation
    [C] Inter-quartile range [D] standard deviation
34. It is true to say that the measure, which is not measure of dispersion, is:
    [A] Range   [B] Variance   [C] Mode   [D] Percentile range
35. The Variance of a given distribution is 25. The standard deviation is:
    [A] 625   [B] 75   [C] 25   [D] 5
36. The standard deviation of the marks 2, 3, 6, 2, 5, 0, 4, 2 is:
    [A] 1.5   [B] 1.7   [C] 1.8   [D] 1.9
37. The standard deviation of the numbers 2, 5, 6, 4 and 8 is:
    [A] 2   [B] 4   [C] 6   [D] 7
38. The mode of the distribution in the table below is:
    [A] 2   [B] 3   [C] 4   [D] 5

Score	0	1	2	3	4	5
Frequency	2	3	4	2	7	2

39. The mean score of the distribution in table above is:
    [A] 1.75   [B] 2   [C] 2.5   [D] 2.75
40. The median score of the distribution in table above is:
    [A] 0   [B] 2.5   [C] 3   [D] 5
41. For a class of 30 students, the scores in a test out of 10 marks were as in the table above. The mode is:
    [A] 3   [B] 5   [C] 6   [D] 7

4	5	7	2	3	6	5	5	8	9
5	4	2	3	7	9	8	7	7	7
3	4	5	5	2	3	6	7	7	2

42. For a class of 30 students, the scores in a test out of 10 marks were as in the table above. The median score is:
    [A] 3 [B] 5 [C] 6 [D] 7
43. For a class of 30 students, the scores in a test out of 10 marks were as in the table above. The range of the distribution is:
    [A] 7 [B] 2 [C] 8 [D] 9
44. The following table shows the tithes in thousand FCFA, collected in a church. The mode is:
    [A] 3 [B] 6 [C] 9 [D] 12

Amount (thousand FCFA)	3	6	9	12	15	18
No. of Christians	3	9	6	15	3	12

45. The table above shows the tithes in thousand FCFA, collected in a church. The median of the distribution is:
    [A] 3 [B] 9 [C] 12 [D] 15
46. The table above shows the frequency distribution of a number of chairs in each rooms of a hotel. The mean of the distribution is:
    [A] 3.5 [B] 4.0 [C] 4.4 [D] 5.0
47. The table above shows the frequency distribution of a number of chairs in each rooms of a hotel. The mode of the distribution is:
    [A] 2 [B] 5 [C] 7 [D] 9
48. Table 37:58, shows the frequency distribution of a number of chairs in each rooms of a hotel. The median of the distribution is:
    [A] 4 [B] 4.5 [C] 5 [D] 5.5
49. The following table shows the frequency distribution of marks scored by a group of students in a test. The number of students who took the test is:
    [A] 14 [B] 15 [C] 18 [D] 20

Marks	2	3	4	5	6
Frequency	2	4	5	3	1

50. The table above shows the frequency distribution of marks scored by a group of students in a test. The modal score is:
    [A] 2 [B] 3 [C] 4 [D] 5
51. The table above shows the frequency distribution of marks scored by a group of students in a test. The mean mark is:
    [A] 1.3 [B] 2 [C] 3 [D] 3.8
52. The following table shows the scores of 15 students in a physics test. The number of students who scored at least 5 is:
    [A] 6 [B] 8 [C] 9 [D] 7

Marks	1	2	3	4	5	6	7	8	9	10
No. of students	1	3	2	0	1	6	1	0	1	0

53. The table above shows the scores of 15 students in a Physics test. The median score is:
    [A] 5 [B] 6 [C] 7 [D] 8

Module 13, Topic 8: Statistics

54. The following table shows the scores of a group of 40 students in a Biology test. If the mode is m and the median is n then (m, n) as an ordered pair is:
    [A] (5,5)   [B] (5,6)   [C] (6,5)   [D] (9,4)

Score	1	2	3	4	5	6	7	8	9
Frequency	2	3	6	7	9	6	2	2	3

55. The table above shows the scores of a group of 40 students in a physics test. The mean of the distribution is:
    [A] 4.5   [B] 4.8   [C] 5.0   [D] 5.2

56. The following table shows the number of goals scored by a football team in 20 matches. The mean number of goals scored is:
    [A] 1.75   [B] 1.9   [C] 2   [D] 2.15

Number of goals	0	1	2	3	4	5
Number of matches	3	5	7	4	1	0

57. The number of goals scored by a football team in 20 matches is shown in the table above. The modal number of goals scored is:
    [A] 1   [B] 2   [C] 5   [D] 7

58. The distribution by Region of 840 students in the faculty of science of the University of Buea in a certain session is as follows:

    Adamawa Region   45
    North West Region   410
    Littoral Region   105
    Western Region   126
    South West Region   154

    In a pie chart drawn to represent this distribution, the angle subtended by Western Region is:
    [A] 42°   [B] 45°   [C] 48°   [D] 54°

59. The pie chart below illustrates the amount of private time a student spends in a week studying various subjects. The value of k is:
    [A] 15°   [B] 30°   [C] 60°   [D] 90°

60. The pie chart in beow illustrates the amount of private time a student spends in a week studying various subjects. Given that, he spends 2 and a half hours on science, the total number of hours he studies in a week is:
    [A] $3\frac{1}{2}$ [B] 5   [C] 8   [D] 12

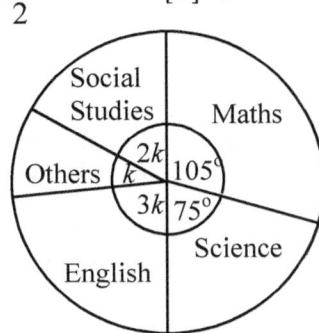

61. The pie chart below represent the number of fruits on display in a grocery shop. Given that there are 60 oranges in display, the number of apples is:

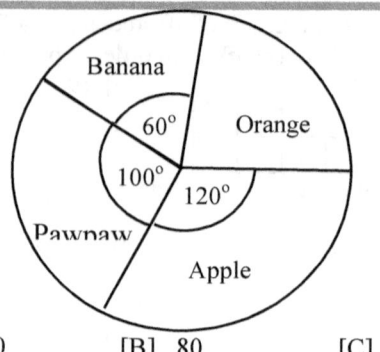

[A] 40 [B] 80 [C] 90 [D] 120

62. The marks obtained by pupils of a certain class are grouped as shown below; 0-4, 5-9, 10-14, 15-19. It is true to say that:
    I: The mid values of the grouped marks are 2,7,12, and 17.
    II: The class interval is 4.
    III: The class boundaries are 0.5, 4.5, 9.5, 14.5 and 19.5.
    [A] I only [B] II only [C] III only [D] I and II

63. The histogram below shows the number of candidates, in thousands who obtained given ranges of marks in an entrance examination.

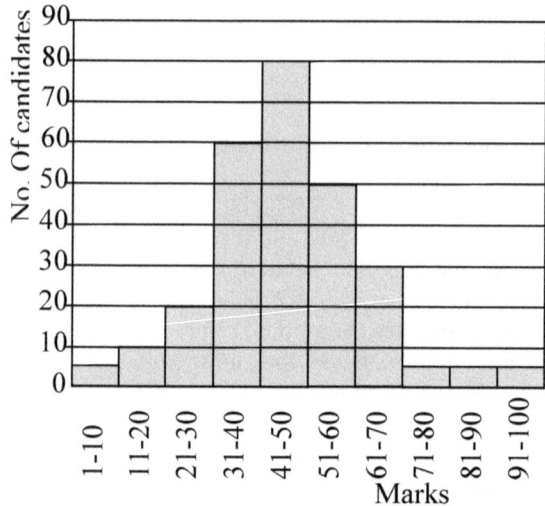

The total number of candidates who sat for the examination is:
    [A] 120,000 [B] 250,000 [C] 260,000 [D] 270,000

64. The histogram above shows the number of candidates, in thousands who obtained given ranges of marks in an entrance examination. The number of candidates who scored at most 30% is:
    [A] 20,000 [B] 25,000 [C] 35,000 [D] 60,000

65. The histogram below shows the distribution of a group of students according to their ages. The range of their ages is:
    [A] 14 years [B] 20 years [C] 30.5 years [D] 31 years

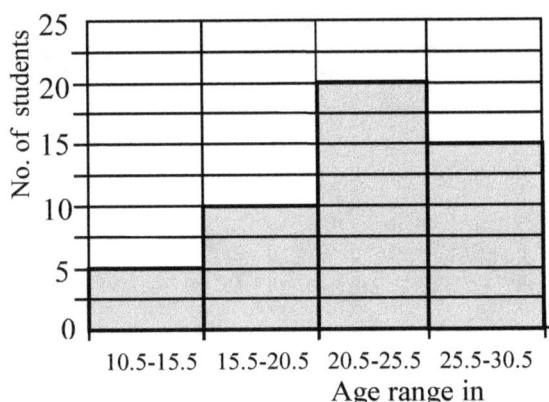

66. The histogram below shows the distribution of a group of students according to their ages. The mode of their ages is:
    [A] 22.5 years   [B] 23.0 years   [C] 24.0 years   [D] 24.5 years
67. For six sequences, Ngange scored 76, 57, 97, 86, 86, 70 in Mathematics. If Ngange wants to convince his parents of his strength in Mathematics, the measure he should use should be:
    [A] Mean   [B] median   [C] mode   [D] range
68. Six employees earn 800 FCFA, 850 FCFA, 900 FCFA, 950 FCFA, 1000 FCFA, and 2350 per hour. The manager claims that the median of the hourly wages is 925 FCFA. The manager is:
    [A] wrong because 925 FCFA is the mode.
    [B] wrong because he seems to ignore the amount 2350 FCFA.
    [C] wrong because 925 FCFA is the mean.
    [D] right because 925 FCFA is the correct median.
69. The president of a certain Credit Union used the data in the table below to find the mean monthly salary of the Credit Union staff.

Monthly Salary	No. of workers
26,000 FCFA	7
30,000 FCFA	8
240,000 FCFA	1
260,000 FCFA	1
300,000 FCFA	3

In a report the president said that, the typical salary at the Credit Union is about 92,000 FCFA. His statement is:
[A] misleading because salaries of five staff are far above those of the other fifteen.
[B] misleading because the mean of the data is not 92,000 FCFA.
[C] misleading because the president ignored the highest salary.
[D] right

Topic 9

PROBABILITY

Objectives

At the end of this topic, the learner should be able to:

1. Define probability or say what probability is about.
2. Understand and the basic probability terminologies such as biased, unbiased, trial, event, outcome, sample space, event subset etc.
3. Calculate simple probabilities.
4. State and use the standard definition of probability.
5. Appreciate that probability is a number in the range $0 \leq P(E) \leq 1$ and that the probability of a sure event is 1, while that of an impossible event is 0.
6. Perform simple experiments using a die and/or playing cards.

Module 13, Topic 9: Probability

9.1 The Concept of Probability

> **? Brainstorming Exercise**
>
> Consider the following statements:
> (i) It will probably rain tonight.
> (ii) It is likely that the principal will address the students tomorrow.
> (iii) The chances of the indomitable lions of Cameroon winning the next world cup football tournament are very high.
>
> 1. How true is each of the above statements?
> 2. What is the likelihood of it raining tonight?
> 3. What is the chance of the indomitable lions wining the next world cup football tournament?
> 4. Questions such as those above lead us into the subject of probability.

Probability seeks to answer questions of chance, likelihood, possibility or the degree of truth in an event or something occurring. In other words, probability is a numerical measure of the degree of chance, possibility or likelihood of an event occurring or not occurring.

We often use probability to take certain life decisions, which depend on chance. For instance, in order to predict the score of say a football march between Egypt and Cameroon, probability comes into play.

9.2 Some Basic Probability Terminology

Suppose a coin is tossed (or thrown), it will either turn up heads (H) or tails (T). If the coin is as likely to turn up heads as to turn up tails, we say then the coin is **fair** or **unbiased**. If the likelihood of the coin turning up heads is not equal to its likelihood of turning up tails we say the coin is **unfair** or **biased**. We call the act of tossing the coin a **trial** or an **experiment** and the appearance of a head or tail an **event**. The event (H or T), one of which must occur in an experiment are called the **outcomes** and the set of all the possible outcomes in a particular experiment is called the **sample space**, usually denoted by S. The sample space S in probability is equivalent to the universal set in set theory. Thus for the case of the coin, $S = (H, T)$

The **event subset** or the **possibility space** is the set of all possible outcomes in an experiment under specified conditions and is a subset of the sample space.

Example

1. An unbiased die is tossed once. Find the sample space and state the event subset if the event is:
   (i) A: obtaining an even number.
   (ii) B: obtaining an odd number.
   (iii) C: obtaining a number less than 3.

   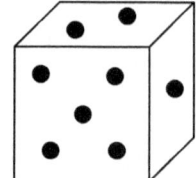

   Solution
   $S = \{1, 2, 3, 4, 5, 6\}$
   (i) $A = \{2, 4, 6\}$
   (ii) $B = \{1, 3, 5\}$
   (iii) $C = \{1, 2\}$

2. A card is drawn at random from a well-shuffled pack of 15 cards numbered 1 to 15. Give the event subsets if the event is obtaining
   (a) X: the number 7.   (b) Y: a number greater than or equal to 10
   (c) Z: a multiple of 3

   Solution
   (a) $X = \{7\}$ (b) $Y = \{10, 11, 12, 13, 14, 15\}$
   (c) $Z = \{3, 6, 9, 12, 15\}$

 Exercise 9:1

1. A student has five pairs of socks of the following colours: blue, red, green, white, and black. On one dark morning, the student chooses a pair of socks at random to put on.
   (a) What is the sample space of this experiment?
   (b) What is the event subset for the event that he chooses a white pair of socks or a green pair of socks?
2. State the set of possible outcome for a simultaneous toss of two fair coins.
3. A die is tossed. What is the sample space?
4. A bag contains 10 tickets numbered 1-10. State the event subset for the event:
   (a) Drawing a prime number,   (b) Drawing an even number,
   (c) Drawing an odd number,   (c) Drawing a square number,
   (e) Drawing a multiple of 3.
5. In an experiment, a teacher chooses a letter at random from the letters of the word 'MATHEMATICS'. What is the possibility space for the event of choosing a vowel?
6. A bag contains 4 red marbles, 7 blue marbles and 1 yellow marble. How many elements, has the sample space?

9.3 Probability as a Number

The probabilities that any event E will occur (or will not occur) always lie between zero and one *inclusively*.

This means that the probability of an event occurring is always zero, one or any number between zero and 1. The probability of an event E occurring is denoted by $P(E)$.

Thus; $0 \leq P(E) \leq 1$

The probability of a **sure** or **certain** event A is 1.

$$P(A) = 1 \Leftrightarrow \text{Event } A \text{ must occur.}$$

For instance, the probability that any living person will die one day is 1.

The probability of an event B occurring is zero, when it is **impossible** for the event to occur.

$$P(A) = 0 \Leftrightarrow \text{Event } A \text{ will not occur.}$$

For instance, the probability that someday a man will be pregnant is zero.

The probability of an event E occurring is any number between 0 and 1 exclusively (not including 0 and 1). This means that there are some chances of the event occurring and some chance of the event not occurring. For instance if it is stated that the probability of an event E occurring is $\frac{1}{4}$ it means that out of 4 trials, the event is expected to occur once and it is expected not to happen 3 times and out of 40 trials, the event is expected to happen 10 times and it is expected not to happen 30 times.

9.4 Equiprobable Outcomes

Equiprobable or **equally likely** events are events, which have equal chances of occurrence. If there are n such events, the probability $P(E)$ of one of the events occurring is given by $P(E) = \frac{1}{n}$.

> **Example**
>
> 1. A fair coin is tossed once. State the probability of:
>   (a) a head   (b) a tail
>
>   **Solution**
> $S = \{H, T\}$
>
>   (a) $P(H) = \dfrac{1}{2}$ (b) $P(T) = \dfrac{1}{2}$
>
> $\Rightarrow P(H) = P(T) = \dfrac{1}{2}$ or 0.5 or 50%
>
> 2. State the probability of each of the faces showing 1, 2, 3, 4, 5, and 6 if a fair die is tossed.
>
>   **Solution**
> $S = \{1, 2, 3, 4, 5, 6\}$
>
> $P(1) = P(2) = P(3) = P(4) = P(5) = P(6) = \dfrac{1}{6}$

9.5 Standard Definition of Probability

Suppose a sample space S consists of a finite number of equiprobable outcomes. Then we define the probability of an event E occurring as

$$\text{Probability of } E = \frac{\text{No of outcomes in the event } E}{\text{Total number of outcomes } S}$$

i.e. $P(E) = \dfrac{n(E)}{n(S)}$

Suits of Playing Cards

An ordinary pack of playing cards contains 52 cards. There are four types of cards; hearts, clubs, diamonds and spades; each type having 13 members labelled A, 1, 2, 3, 4, 5, 6, 7, 8, 9, 10, Q, K, and J. Each type of card has 3 picture cards labeled **Q**, **K** and **J,** called queen, king or jack. Hearts and diamonds are red while clubs and spades are black.

Module 13, Topic 9: Probability

Ace of hearts   Ace of clubs   Ace of diamonds   Ace of spades

Jack of hearts   Queen of spades   King of diamonds   Jack of spades

 Example

1. A boy picks a card at random from a well-shuffled pack of 52 playing cards. What is the probability that (i) it is an Ace of heart (ii) it is a king

   Solution
   $n(S) = 52$
   (i) $n(\text{Ace of hearts}) = 1$
   $$\therefore P(\text{Ace of hearts}) = \frac{n(\text{Ace of hearts})}{n(S)} = \frac{1}{52}$$
   (ii) $n(\text{Kings}) = 4$
   $$\therefore P(\text{Kings}) = \frac{n(\text{Kings})}{n(S)} = \frac{4}{52} = \frac{1}{13}$$

2. Fourteen girls are sitting in a circle equally spaced. One is from form four, 2 are from form five, 6 are from lower sixth and 5 are from upper sixth. A girl is selected at random from amongst the girls. Find the probability that the girl is from
   (i) form four   (ii) form five   (iii) lower sixth   (iv) upper sixth

   Solution
   $n(S) = 14$, $n(\text{form four}) = 1$, $n(\text{form five}) = 2$
   $n(\text{Lower } 6^{th}) = 6$ and $n(\text{upper } 6^{th}) = 5$

(i) $P(\text{form four}) = \dfrac{n(\text{form four})}{n(s)} = \dfrac{1}{14}$

(ii) $P(\text{form five}) = \dfrac{n(\text{form five})}{n(5)} = \dfrac{2}{14} = \dfrac{1}{7}$

(iii) $P(\text{lower } 6^{th}) = \dfrac{n(\text{lower } 6^{th})}{n(s)} = \dfrac{6}{14} = \dfrac{3}{7}$

(iv) $P(\text{upper } 6^{th}) = \dfrac{n(\text{upper } 6^{th})}{n(s)} = \dfrac{5}{14}$

Exercise 9:2

1. A die is tossed. Find the probability of:
   - (a) obtaining a 2.
   - (b) obtaining an even number.
   - (c) obtaining an odd number.
   - (d) obtaining a number less than 5.
   - (e) obtaining a prime number.

2. The figure below shows a spinner. Find the probability of obtaining:
   - (a) 3   (b) 1   (c) 2   (d) 5

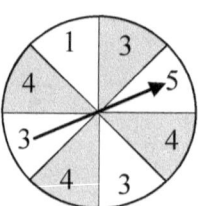

3. In a race of twenty horses, 7 of the horses are black, 8 are white and the rest are dotted. Find the probability that the winner will be dotted.

4. At a conference, there are 9 boys, 12 girls, 15 men and 14 women. Find the probability that a person elected as president will be a man.

5. Find the probability of choosing at random the letter B from the letters of the word 'PROBABILITY'.

6. Find the probability of choosing a prime number at random from the integers 5 to 25 inclusive.

Module 13, Topic 9: Probability

Multiple Choice Exercise 9

1. The probability of having an odd number in a single toss of a fair die is:

   [A] $\dfrac{2}{3}$ [B] $\dfrac{1}{6}$ [C] $\dfrac{1}{3}$ [D] $\dfrac{1}{2}$

2. 11. A number is chosen at random from the set $\{1, 2, 3, \cdots, 9, 10\}$. The probability that the number is greater than or equal to seven is:

   [A] $\dfrac{1}{10}$ [B] $\dfrac{3}{10}$ [C] $\dfrac{2}{5}$ [D] $\dfrac{3}{5}$

3. The probability of throwing a number greater than 2 with a single fair die is:

   [A] $\dfrac{1}{6}$ [B] $\dfrac{1}{3}$ [C] $\dfrac{1}{2}$ [D] $\dfrac{2}{3}$

4. The probability of obtaining 4 or 6 on rolling a fair die once is:

   [A] $\dfrac{2}{3}$ [B] $\dfrac{1}{6}$ [C] $\dfrac{1}{3}$ [D] $\dfrac{1}{2}$

5. Three balls are drawn one after the other with replacement, from a bag containing five red, nine white and four blue identical balls. The probability that they are one red, one white and one blue is:

   [A] $\dfrac{5}{81}$ [B] $\dfrac{5}{27}$ [C] $\dfrac{5}{162}$ [D] $\dfrac{5}{243}$

6. The probability that an integer selected from the set of integers $\{20, 21, \cdots, 30\}$ is a prime number is:

   [A] $\dfrac{2}{11}$ [B] $\dfrac{5}{11}$ [C] $\dfrac{6}{11}$ [D] $\dfrac{9}{11}$

7. The probability of obtaining a number less than three is on rolling a fair die once is:

   [A] $\dfrac{1}{6}$ [B] $\dfrac{1}{3}$ [C] $\dfrac{1}{2}$ [D] $\dfrac{2}{3}$

8. The probability that a number selected from the numbers 30 to 50 inclusive is a prime is:

   [A] $\dfrac{1}{4}$ [B] $\dfrac{5}{21}$ [C] $\dfrac{3}{7}$ [D] $\dfrac{1}{3}$

9. The probability of selecting a prime number at random from the set $Y = \{18, 19, 20, \cdots, 28, 29\}$ is:

   [A] $\dfrac{1}{4}$ [B] $\dfrac{3}{11}$ [C] $\dfrac{1}{2}$ [D] $\dfrac{3}{4}$

Module 14

Algebra and Logic

Family of Situations
At the end of the module 14; the student is expected to acquire many more competencies within the **families of situations** *'Describing patterns and relationships between quantities using symbols'*.

Categories of Action
The categories of action for module 14 include:
1. Interpretation of algebraic models,
2. Determination of quantities from algebraic models
3. Representation of quantities and relationships.

Credit
The module is expected to be covered within 10 weeks teaching 4 periods of 50 minutes per week (or within 40 periods).

Topic 10

SIMPLE ALGEBRA

Objectives
At the end of this topic, the learner should be able to:

1. Expand $(a \pm b)^2$ and $(a+b)(a-b)$.
2. Factorize 4 terms expressions.
3. Factorize expressions of the form $a^2 - b^2$ and $ax^2 + bx + c$ with $a \neq 0$.

10.1 Review of Algebraic Expressions

 Review

1. Translate the following into algebraic expressions.
   (a) The square of a number increased by two.
   (b) Two times a number decreased by eight.
   (c) Seven less than half a number.
   (d) Half a number increased by seven.
2. If $y = -1$ and $x = 4$, find the value of $x + 3y$.
3. The formula for finding the volume, V of a cylinder is $V = \pi r^2 h$. Find the volume of a cylinder whose radius r is 2 cm and whose height h is 7 cm. Take $\pi = \frac{22}{7}$.
4. (a) How many terms has the expression $6x - 2xy + 3y - 5$?
   (b) Write down the terms of this expression.
5. Evaluate the following.   (a) $8x - 4y + 13x - 13y$
   (b) $6s + 7t - 8t - 5s$ (c) $12p - 9q - 13q - 7p$
6. Simplify the following.

   (a) $3a(10a)$ (b) $5x(4xy)$ (c) $28xy \div 4$ (d) $\dfrac{7x^2 y}{8x}$

7. Find the HCF and LCM of $24x^2 y$ and xy^2.
8. Expand the following.
   (a) $3(x+5)$ (b) $(y+2)4$ (c) $x(4+y)$ (d) $(3-u)v$

10.2 Expansion of a Product of Two Binomials

Using the idea of the expansion of the product of a monomial and a binomial developed in topic 14 of form 2, we can expand the product of two binomials as follows.

$$(a+b)(c+d) = a(c+d) + b(c+d) = ac + ad + bc + bd$$

$$\therefore \quad (a+b)(c+d) = ac + bc + ad + bd$$
$$(a+b)(c-d) = ac + bc - ad + bd$$
$$(a-b)(c-d) = ac - bc - ad + bd$$

 Example

Expand (a) $(x+1)(x+2)$ (b) $(y-2)(y+3)$

Solution

(a) $(x+1)(x+2) = x(x+2)+1(x+2)$
$\qquad = x^2+2x+x+2$
$\qquad = x^2+3x+2$

(b) $(y-2)(y+3) = y(y+3)-2(y+3)$
$\qquad = y^2+3y-2y-6$
$\qquad = y^2+y-6$

10.3 Expansion of the Square of a Binomial

$(a+b)^2 = (a+b)(a+b) = a(a+b)+b(a+b)$
$\qquad = a^2+ab+ab+b^2 = a^2+2ab+b^2$

$\boxed{(a+b)^2 = a^2+2ab+b^2}$ ....①

(b) $(a-b)^2 = (a-b)(a-b) = a(a-b)-b(a-b)$
$\qquad = a^2-ab-ab+b^2 = a^2-2ab+b^2$

$\boxed{(a-b)^2 = a^2-2ab+b^2}$ ....②

The identities ① and ② are examples of **perfect squares**. A perfect square is a number or expression, which has an exact square root. Numerical examples of perfect squares are 1, 4, 9, 16, 25, 36, 49, etc. The above expansions are very useful in computing numerical perfect squares.

Example

Evaluate (a) 17^2 (b) 99^2

Solution
(a) $17^2 = (10+7)(10+7) = 100+140+49 = 289$
(b) $99^2 = (100-1)(100-1) = 10000-200+1 = 9801$

Exercise 10:1

(a) Expand the following.
1. $3(x+5)$
2. $(y+2)4$
3. $x(4+y)$
4. $(3+u)v$
5. $2(x-1)$
6. $(y-3)5$
7. $3p(2-q)$
8. $(4-s)t$
9. $2x(5-x)$
10. $(x+1)(x+3)$
11. $(x-1)(x+2)$

12. $(p+3)(p-1)$ 13. $(x+3)^2$ 14. $(x-2)^2$
15. $(4+a)^2$ 16. $(3-y)^2$

(b) Evaluate the following.
   (1) 49^2 (2) 101^2 (3) 198^2 (4) 442^2

10.4 Simple Factorization

Factorization is the reverse process of expansion. To factorize, extract the HCF of all the terms. For instance, since $a(b+c) = ab+ac$, a is the HCF of ab and ac. So to factorize $ab+ac$, remove a from the bracket. Thus,

$$ab + ac = a(b+c)$$

 Example

Factorise the following:

(a) $2x + 2y$ (b) $4xy + 12xz$ (c) $\frac{1}{4}pqr - \frac{5}{12}pqs$

Solution

(a) Since HCF of $2x$ and $2y$ is 2,
    $2x + 2y = 2(x+y)$

(b) Since the HCF of $4xy$ and $12xz$ is $4x$,
    $4xy + 12xz = 4x(y + 3z)$

(c) Since HCF of $\frac{1}{4}pqr$ and $-\frac{5}{12}pqs$ is $\frac{1}{4}pq$,
    $\frac{1}{4}pqr - \frac{5}{12}pqs = \frac{1}{4}pq\left(r + \frac{5}{3}s\right)$

 Exercise 10:2

Factorise:
1. $3x + xy$ 2. $2p + 6q$ 3. $5x - 10y$ 4. $4xy + 8y$
5. $6uv - 3uf$ 6. $\frac{1}{4}xy + \frac{1}{4}px$ 7. $\frac{1}{2}ax - \frac{1}{2}bx$ 8. $\frac{2}{3}y - \frac{1}{3}x$

Module 14, Topic 10: Simple Algebra

10.5 Factorization by Grouping

In factorizing more than three terms, group similar terms.

Example

Factorize the following:
(a) $px - py + qx - qy$ (b) $x - y + xy - 1$ (c) $6x - 6y + 3ax - 3ay$

Solution
(a) $px - py + qx - qy = p(x-y) + q(x-y) = (x-y)(p+q)$
(b) Rearrange
$$x - y + xy - 1 = xy - y + x - 1$$
$$= (xy - y) + (x - 1)$$
$$= y(x - 1) + 1(x - 1)$$
$$= (x - 1)(y + 1)$$
(c) $6x - 6y + 3ax - 3ay = (6x - 6y) + (3ax - 3ay)$
$$= 6(x - y) + 3a(x - y)$$
$$= (x - y)(6 + 3a)$$

Exercise 10:3

Factorise
(a) $xy + x + y + 1$
(b) $ax + a + 3x + 3$
(c) $6px + 4p + 3x + 2$
(d) $3x - 6xy + 2 - 4y$
(e) $6y - 9x - 4y + 6$
(f) $5u - ut - 5v + tv$
(g) $an + am - 3m - 3n$
(h) $pr + 3ps - 2qr - 6qs$
(i) $x^2 - 4x + 6xy - 24y$
(j) $2q^3 - 14q^2 + 3q - 21$
(k) $24x + 2ab - 3bx - 16a$
(l) $p^2 - 5 - p^2q + +5q$

10.6 The Difference of Two Squares

Consider the expansion $(a+b)(a-b) = a^2 - b^2$.

Competency Based Mathematics for Secondary Schools. Book 3

We call an expression such as $a^2 - b^2$ the **difference of two squares**. The difference of two squares is very useful in evaluating numerical expressions such as $19^2 - 13^2$, $925^2 - 725^2$

Example

1. Using the idea of the difference of two squares, evaluate the following.
   (a) $4^2 - 3^2$ (b) $25^2 - 16^2$ (c) $19^2 - 13^2$ (d) $925^2 - 725^2$

   Solution
   (a) $4^2 - 3^2 = (4-3)(4+3) = 1(7) = 7$
   (b) $25^2 - 16^2 = (25-16)(25+16) = 9(41) = 369$
   (c) $19^2 - 13^2 = (19-13)(19+13) = 6(32) = 192$
   (d) $925^2 - 725^2 = (925-725)(925+725) = 200(1650) = 330,000$

2. Factorize the following:
   (a) $x^2 - y^2$ (b) $(ax)^2 - (by)^2$ (c) $25a^2 - 9b^2$ (d) $a^2 - b^2c^2$

   Solution
   (a) $x^2 - y^2 = (x-y)(x+y)$
   (b) $(ax)^2 - (by)^2 = (ax-by)(ax+by)$
   (c) $25a^2 - 9b^2 = (5a)^2 - (3b)^2 = (5a-3b)(5a+3b)$
   (d) $a^2 - b^2c^2 = a^2 - (bc)^2 = (a-bc)(a+bc)$

Exercise 10:4

1. Factorize:
   (a) $x^2 - y^2$ (b) $1 - 9y^2$ (c) $36 - 9x^2$ (d) $4a^2 - 1$
   (e) $4y^2 - 9x^2$ (f) $4u^2 - 25y^2$ (g) $36a^2 - 49b^2$ (h) $x^2 - y^2z^2$

2. Evaluate the following:
   (a) $41^2 - 40^2$ (b) $124^2 - 120^2$ (c) $625^2 - 525^2$
   (d) $3.8^2 - 3.7^2$ (e) $1375^2 - 1325^2$ (f) $0.003^2 - 0.002^2$

Module 14, Topic 10: Simple Algebra

10.7 Factorizing Quadratic Expressions

Consider the expansion
$$(2x + 1)(x + 2) = 2x^2 + 5x + 2$$
The right hand side of this identity is of the form $ax^2 + bx + c$, where a, b, c are constants and $a \neq 0$. This is the general form of a quadratic expression. The power of the unknown (x) is 2.

To factorize a quadratic expression, we should take the following steps.

(i) Multiply a by c to have ac
(ii) Find the pair of integral factors p and q of ac whose sum or difference is b.
(iii) Substitute the middle term bx with the sum or difference of px and qx in $ax^2 + bx + c$.
(iv) Factorize the expression by grouping.

 Example

Factorize:
1) $x^2 + 7x - 3$ 2) $2x^2 - x - 1$ 3) $12x^2 + 8x - 15$

Solutions

1) $6x^2 + 7x - 3$

   $ac = (6)(-3) = -18$ and $b = 7$
   Pairs of factors of 18 are {1, 18}, {2, 9}, {3, 6}.
   Since $9 - 2 = 7$, it means -2 and 9 are the required factors.
   $\Leftrightarrow 6x^2 + 7x - 3 = 6x^2 + 9x - 2x - 3$
   Factorizing by grouping
   $6x^2 + 7x - 3 = (6x^2 + 9x) + (-2x - 3)$
   $= 3x(2x + 3) - 1(2x + 3)$
   $= (2x + 3)(3x - 1)$

2) $2x^2 - x - 1$

   $ac = -2$
   Pair of factors of $2 = \{1, 2\}$ and $-2 + 1 = -1$
   $2x^2 - x - 1 = 2x^2 - 2x + x - 1$
   Factorizing by grouping
   $2x^2 - x - 1 = (2x^2 - 2x) + (x - 1) = 2x(x - 1) + 1(x - 1) = (x - 1)(2x + 1)$

3) $12x^2 + 8x - 15 = 12x^2 + 18x - 10x - 15$
   $= 6x(2x + 3) - 5(2x + 3) = (2x + 3)(6x - 5)$

When a quadratic is factorable, a binomial factor of the first bracket during grouping is always certainly a factor of the second bracket. For instance in Example (3) above $2x + 3$ is a binomial factor of the first and second bracket of the expression $6x(2x + 3) - 5(2x + 3)$.

Please always take advantage this fact.

When the coefficient of x^2 is unity, i.e. $a = 1$, the factorization becomes simpler.

 Example

1. Factorize:
    (1) $x^2 + x - 2$ (2) $x^2 + 9x + 18)$
    (3) $3x^2 - 12x - 63$

Solution
(1) Factors of –2 whose sum is +1 are, +2 and –1
$$x^2 + x - 2 = (x+2)(x-1)$$
(2) Factors of +18 whose sum is +9 are, +6 and +3
$$x^2 + 9x + 18 = (x+6)(x+3)$$
(3) $3x^2 - 12x - 63 = 3\{x^2 - 4x - 21\}$
Factors of –21 whose sum is – 4 are, –7 and +3.
$$\Rightarrow 3x^2 - 12x - 63 = 3\{(x-7)(x+3)\}$$

 Exercise 10:5

Factorise the following.
(1) $x^2 + x - 6$ (2) $p^2 + 12p + 11$ (3) $y^2 - 7y + 12$
(4) $x^2 + 6x - 16$ (5) $x^2 - 2x - 15$ (6) $10x^2 - x - 3$
(7) $4a^2 - 3a - 10$ (8) $5 - 16y + 12y^2$ (9) $3 - x - 2x^2$
(10) $10 - 11x + 3x^2$

Module 14, Topic 10: Simple Algebra

 Multiple Choice Exercise 10

1. The expression, which is a perfect square, is:
    [A] $(x+1)(x-1)$ [B] x^2+2x+1 [C] x^2-y^2 [D] x^2-2x-1
2. The expression, which is not a difference of two squares, is:
    [A] $(x+1)(x-1)$ [B] x^2-1 [C] x^2-y^2 [D] $(x-y)^2=6$
3. An identity among the following is:
    [A] $(x+1)(2x+3)=2x^2+5x+3$ [B] $(3p-1)(2p+1)=3p^2-5p-1$
    [C] $2y+7=3y-5$ [D] $x^2+2x+1=6$
4. The statement, which is not an identity, is:
    [A] $(x+2)(x-1)=x^2+x-2$ [B] $(5x-1)(x+1)=5x^2+4x-1$
    [C] $(3x+1)(2x-1)=6x^2-x-1$ [D] $(3x-1)(2x+1)=3x^2-5x-1$
5. When simplified, $5yx-7xy+4yx$ equals:
    [A] $9yx$ [B] $-9xy$ [C] $8xy$ [D] $2xy$
6. The simplified form of $6p+7q-8q-5p$ is:
    [A] $p-q$ [B] $q-p$ [C] $p+q$ [D] $11p-5q$
7. We can simplify $-7x+8y-2+9x-10y+4$ to have:
    [A] $2(x-y+1)$ [B] $2(x-y-1)$ [C] $2(x+y-1)$ [D] $2(x+y+1)$
8. Simplifying $15x-12y+8z-14x+12y-8z$ leads to:
    [A] x [B] y [C] z [D] $x-y+z$
9. By simplification $x+(-x)+y$ is exactly:
    [A] $-y$ [B] y [C] $2x-y$ [D] $2x+y$
10. $32e+6f-12e+4f$ can also be:
    [A] $38e+6f$ [B] $30e+8f$ [C] $28e+9f$ [D] $20e+10f$
11. When expanded $-2a(3a^2b+4b^2)$ gives:
    [A] $-6ab^2-8a^2b$ [B] $-6ab^2-4ab^2$
    [C] $-6ab^2+8a^2b$ [D] $-6a^3b-8ab^2$
12. On Simplification $13x-(2x-4x-3x)$ becomes:
    [A] $8x$ [B] $18x$ [C] $-8x$ [D] $-18x$
13. $9x-(5x-3y)-y$ is equal to:
    [A] $4x-2y$ [B] $4x+2y$ [C] $5x-2y$ [D] $5x+2y$
14. $-2a-5b-(8b-5a)=$
    [A] $-8a+13b$ [B] $3a+3b$ [C] $3a-13b$ [D] $7a-13b$
15. $(2x+y)+(x-2y)$ is the same as:
    [A] $3x+y$ [B] $x-3y$ [C] $x+3y$ [D] $3x-y$
16. $(2x+y)-(x-2y)$ is equal to:

[A] $3x - y$ [B] $x + 3y$ [C] $x - 3y$ [D] $3x + y$

17. $(2x - 3) - (2 - 3x)$ is equal to:
    [A] $5x - 5$ [B] $5x - 1$ [C] $x - 5$ [D] $x - 1$

18. Adding $(2x + y)$ and $(x - 2y)$ gives:
    [A] $3x + y$ [B] $x - 3y$ [C] $x + 3y$ [D] $3x - y$

19. Given the statement $x - 13y + 5z - 4m = x - ($ $)$
    The expression required in the bracket is:
    [A] $-13y + 5z - 4m$ [B] $-13y + 5z$
    [C] $-13y + 5z - 4x$ [D] $13y - 5z + 4m$

20. On expansion $(2x - 5)(x - 3)$ gives:
    [A] $x^2 - 11x - 15$ [B] $2x^2 - 11x + 15$
    [C] $2x^2 - 5x - 8$ [D] $x^2 - 5x + 15$

21. Given that $p = 3 - 2y$ and $q = 4 + 3y$. The value of pq is:
    [A] $-6y^2 - y - 12$ [B] $6y^2 - y - 12$ [C] $-12 + y + 6y^2$ [D] $12 + y - 6y^2$

23. $(2x - 1)(x + 2)$ is equal to:
    [A] $2x^2 - 2$ [B] $2x^2 + x - 2$ [C] $2x^2 - x - 2$ [D] $2x^2 - 3x - 2$

25. The product of $x - 1$ and $x + 1$ is:
    [A] 2 [B] $2x$ [C] $x^2 + 2x - 1$ [D] $x^2 - 1$

26. If $(a + b)^2 = a^2 + 2ab + b^2$ the value of $(2a + 1)^2$ is:
    [A] $4a^2 + 4a - 1$ [B] $4a^2 + 4a + 1$ [C] $4a^2 - 4a - 1$ [D] $4a^2 - 4a + 1$

27. The square of $x - 8$ is equal to:
    [A] $x^2 - 16x - 64$ [B] $x^2 + 16x - 64$
    [C] $x^2 - 16x + 64$ [D] $x^2 - 32x + 64$

28. The coefficient of x in the expansion of $(x + 9)(x + 3)$ is:
    [A] -12 [B] 12 [C] 3 [D] -3

29. The coefficients of x and x^2 in the expansion of $(x-3)^2$ are respectively:
    [A] $-6, 1$ [B] $6, 1$ [C] $-1, 6$ [D] $1, -6$

30. The coefficient of xy in the expansion of $(3x + 2y)(4x - 2y)$ is:
    [A] -2 [B] -14 [C] 2 [D] 14

31. When factorized $3x(4 - y) - m(y - 4)$ becomes:
    [A] $(3x + m)(4 - y)$ [B] $(3x - m)(4 - y)$
    [C] $(3x + m)(y - 4)$ [D] $(3x - m)(y + 4)$

32. We can factorize the expression $x(a - c) + y(c - a)$ to obtain:
    [A] $(a - c)(y + x)$ [B] $(a - c)(x - y)$
    [C] $(a + c)(x - y)$ [D] $(a - c)(y - x)$

Module 14, Topic 10: *Simple Algebra*

33. By factorizing $m(2a - b) - 2n(b - 2a)$, the result is:
    [A] $(2a - b)(2n - m)$ [B] $(2a - b)(m - 2n)$
    [C] $(2a - b)(m + 2n)$ [D] $(2a - b)(m - 2n)$
34. The difference between the squares of the numbers 21 and 11 is:
    [A] 20   [B] 100   [C] 220   [D] 320
35. The value of $13^2 - 12^2$ is:
    [A] 25   [B] 5   [C] 1^2 [D] 125
36. $32x^3 - 8xy^2$ when factorized gives:
    [A] $4(4x + y)(2x - y)$ [B] $(16x - y)(2x + y)$
    [C] $8x(2x - y)$ [D] $8x(2x + y)(2x - y)$
37. By factorizing $27p^2x^2 - 48y^2$ the result is:
    [A] $3(3px - 4y)(3px + 4y)$ [B] $9(3px - 4y)^2$
    [C] $9(px - 4y)(3px + 4y)$ [D] $3(3px - 4y)^2$
38. $(x - 2)(x + 3)$ are the factors of:
    [A] $x^2 - 9$ [B] $x^2 - 6$ [C] $x^2 - x - 6$ [D] $x^2 + x - 6$
39. The result of factorizing $x^2 + 4x - 192$ is:
    [A] $(x - 4)(x + 48)$ [B] $(x + 48)(x + 4)$
    [C] $(x - 12)(x + 16)$ [D] $(x - 12)(x - 16)$
40. When factorized, the expression $2x^2 + x - 15$ equals:
    [A] $(2x + 5)(x - 3)$ [B] $(2x - 5)(x + 3)$
    [C] $(2x - 5)(x - 3)$ [D] $(2x - 3)(x + 5)$
41. The quadratic $2e^2 - 3e + 1$ when factorised, becomes:
    [A] $(2e - 1)(e - 1)$ [B] $(e^2 - 3)(2e - 1)$
    [C] $(2e + 3)(e - 2)$ [D] $(2e - 3)(e - 1)$
42. Factorising $3a^2 - 11a + 6$ leads to:
    [A] $(3a - 2)(a - 3)$ [B] $(2a - 2)(a - 3)$
    [C] $(3a - 2)(a + 3)$ [D] $(3a + 2)(a - 3)$
43. $2x^2 - 9x - 45$, written as the product of two factors is:
    [A] $(2x - 9)(x - 5)$ [B] $(2x - 15)(x + 3)$
    [C] $(2x + 15)(x - 3)$ [D] $(2x - 15)(x - 3)$
44. On factorisation $6x^2 + 7x - 20$ becomes:
    [A] $(6x - 5)(x + 4)$ [B] $2(3x - 5)(x + 2)$
    [C] $(3x + 4)(2x - 5)$ [D] $(3x - 4)(2x + 5)$
45. We can factorise the quadratic $3p^2 + 2p - 1$ to have:
    [A] $(3p + 2)(p - 1)$ [B] $(3p - 1)(p + 1)$
    [C] $(3p + 1)(p - 1$ [D] $(3p - 2)(p + 1)$
48. The result of factorising $mn - xy - nx + my$ is:
    [A] $(n + y)(m - x)$ [B] $(n - y)(m + x)$
    [C] $(x - m)(n - y)$ [D] $(n - y)(m - x)$

Topic 11

EQUATIONS

Objectives
At the end of this topic, the learner should be able to:

1. Solve multi-step simple linear equations.
2. Solve simultaneous linear equations in two equations using the method of elimination and substitution.
3. Solve quadratic equations using the factorizations method and formula method.
4. Solve real life problems that lead to simple linear equations, linear, simultaneous linear equations or quadratic equations.

Module 14, Topic 11: Equations

Much work was done in topic 16 of book 2 on simple linear equations. We shall review that work before continuing to other types of equations; notably simultaneous equations and quadratic equations. It is very important for you to go back and revise that section before continuing.

11.1 Simple Linear Equations

 Review

Solve the following equations.

1. $4x = 32$
2. $15 = 5t$
3. $-3p = 18$
4. $\dfrac{u}{7} = 6$
5. $4 = \dfrac{z}{9}$
6. $\dfrac{w}{-3} = 11$
7. $-9 = \dfrac{r}{3}$
8. $\dfrac{-5q}{2} = 15$
10. $5u = 2u + 27$
11. $5x = 40 - 3x$
12. $2x = 90 - 7x$
13. $10t - 11 = 8t$
14. $18 - 5a = a$
15. $9u = 16u - 105$
16. $5p + 3 - 2p = p + 8$
17. $7m + 10 - 2m = 16m - 12$
18. $-6p - 21 = 3p - 12$
19. $3x + 5 = 21 - x$
20. $p - 0.1p + 0.9 = 0.2(p + 1)$
21. $100 + 3\tfrac{1}{2}x = 23\tfrac{1}{2}x$

22. MTN charges 250 Frs. per minute for calls. After making an MTN call that last m minutes, NTN charges Konyuy a bill of 1750 Frs. For how many minutes did Konyuy make the call?
23. Mbianda bought a radio that costs 16960 FCFA. With tax, t she pays 17810 FCFA. Find the value of t.
24. Tanto intends to buy a bicycle, which costs 27000 FCFA. He has 6000 FCFA and saves 3000 FCFA every week for n weeks. Write down an equation and use it to calculate the number of weeks he saves 3000 FCFA.

11.2 Simultaneous Linear Equations

 Group Activity

A tailor bought 2 razors and 3 needles at 70 FRS and a seamstress bought 3 razors and 5 needles at 110 FRS from the same store.
You need 4 razors and 7 needles. How much would you budget for these items?

This very simple practical real life problem occurs almost on a daily bases. People who have studied simultaneous linear equations will very easily recognize these types of problems and tackle them with a pinch of salt. However if you don't know how to solve simultaneous linear equations you cannot just attempt the solution.
Simultaneous linear equations are a pair of linear equations in two unknowns

which can satisfy both equations simultaneously.

An example of simultaneous linear equations is,

$$7x + y = 22$$
$$5x + y = 14$$

Before attempting the above problem, lets first study how to solve simultaneous linear equations.

11.3 Simultaneous Equations with Uniform Coefficients

Method of Elimination

In this method, we add or subtract the equations to eliminate one of the unknowns. After eliminating one unknown, the remaining equation is a simple linear equation. We now solve for the unknown in this simple linear equation and then substitute into any of the original equations to find the other unknown.

 Example

Solve the following simultaneous equations using the method of elimination.
(a) $7x + y = 22$ (b) $3a - b = 21$
      $5x + y = 14$ $2a + b = 4$

(c) $5m + 3n = 17$ (d) $4p - 2q = -14$
      $m + 3n = 1$ $4p - 5q = -32$

Solutions
(a) $7x + y = 22$ ....①
      $5x + y = 14$ ....②
Subtract equation ② from equation ①
$$2x = 8 \Rightarrow x = 4$$
Substitute in equation ②
$$5(4) + y = 14$$
$$20 + y = 14$$
Subtract 20 from both sides
$$\Rightarrow y = -6$$
(b) $3a - b = 21$ ....①
      $2a + b = 4$ ....②

Add equation ① to equation ②.
$5a = 25$ and $a = 5$
Substitute in equation ②
$$2(5) + b = 4$$
$$10 + b = 4 \Rightarrow b = -6$$

Alternatively $a = 5$ could have been substituted in equation ① instead of equation ② as follows.
$$3(5) - b = 21$$
$$-b = 6 \Rightarrow b = -6$$

(c) $5m + 3n = 17$ .... ①
$m + 3n = 1$ .... ②
① − ②: $4m = 16 \Rightarrow m = 4$
Substitute in ②
$$4 + 3n = 1$$
$$3n = -3 \Rightarrow n = -1$$

(d) $4p - 5q = -32$ .... ①
$4p - 2q = -14$ .... ②
② − ① : $3q = 18 \Rightarrow q = 6$
Substitute in equation ②
$$4p - 2(6) = -14$$
$$4p = -2 \Rightarrow p = -\frac{1}{2}$$

Exercise 11:1

Solve the following simultaneous equations using the method of elimination.

1. $x + y = 3$
   $3x - y = 1$

2. $y + x = 4$
   $2y - x = 5$

3. $a = 2b - 1$
   $2a = 2b + 1$

4. $x + y = 4$
   $x - y = 2$

5. $2x - 3y = 5$
   $3x + 2y = 14$

6. $6x + 5y = 11$
   $5x - 2y = 5$

7. $7x + 3y = 6$
   $5y - 9x = 10$

8. $y = 4x$
   $3x + y = 21$

Competency Based Mathematics for Secondary Schools. Book 3

Method of Substitution

To solve simultaneous linear equations using the method of substitution, first solve one of the equations for one of the unknowns. Then substitute the result into the other equation to find the value of the other unknown. Substitute the value of this unknown into the subject equation of the other unknown to find it.

 Example

Solve the following simultaneous equations using the method of substitution.

(a) $7x + y = 22$
$5x + y = 14$

(b) $3a - b = 21$
$2a + b = 4$

(c) $5m + 3n = 17$
$m + 3n = 1$

(d) $4p - 2q = -14$
$4p - 5q = -32$

(a) $7x + y = 22$ ....①
$5x + y = 14$ ....②
From equation ②
$y = 14 - 5x$ ....③
Substitute in equation ①
$7x + 14 - 5x = 22$
$2x + 14 = 22$
$2x = 8 \Rightarrow x = 4$
Substitute in equation ③
$y = 14 - 5(4) \Rightarrow y = -6$

(b) $3a - b = 21$ ....①
$2a + b = 4$ ....②
From equation ②
$b = 4 - 2a$ ....③
Substitute in equation ①
$3a - (4 - 2a) = 21$
$5a - 4 = 21$
$5a = 25 \Rightarrow a = 5$
Substitute in equation ③
$b = 4 - 2(5) \Rightarrow b = -6$

(c) $5m + 3n = 17$ ....①
$m + 3n = 1$ ....②

From equation ②,
$$m = 1 - 3n \quad\quad\quad\quad\quad ③$$
Substitute in equation ①
$$5(1-3n) + 3n = 17$$
$$5 - 15n + 3n = 17$$
$$-15n + 3n = 12$$
$$-12n = 12 \Rightarrow n = -1$$
Substitute in equation ③
$$m = 1 - 3(-1) \Rightarrow m = 4$$

(d) $4p - 5q = -32 \quad\quad\quad\quad ①$
$4p - 2q = -14 \quad\quad\quad\quad ②$
From equation ②
$$2p - q = -7$$
$$-q = -7 - 2p$$
$$q = 7 + 2p \quad\quad\quad\quad ③$$
Substitute in equation ①
$$4p - 5(7 + 2p) = -32$$
$$4p - 35 - 10p = -32$$
$$-6p = 3 \Rightarrow p = -\frac{1}{2}$$
Substitute in equation ③
$$q = 7 + 2\left(-\frac{1}{2}\right) \Rightarrow q = 6$$

Notice that though we can solve any simultaneous linear equation in two unknowns using either method, *simultaneous equations with uniform coefficients are easier to solve using the method of elimination.*

Exercise 11:2

Solve the simultaneous equations in Exercise 11:1 above using the method of substitution.

11.4 Simultaneous Equations with Non-uniform Coefficients

Method of Elimination

To solve these types of simultaneous equations by elimination, the objective is to eliminate one of the unknowns. To do this, make the coefficients of one of the unknowns' uniform by multiplying both equations by factors of the LCM of the coefficients of one of the unknowns. In this way, the LCM will be the uniform coefficient. In order that the problem does not become very complicated with large numbers, it is advisable that the LCM should be that of the smaller coefficients.

Example

Solve the following simultaneous equations using the method of elimination.

(a) $x + 2y = 5$
$3x - y = 1$

(b) $2x + 3y = 7$
$5x - 2y = 8$

Solution

(a) $x + 2y = 5$ ....①
$3x - y = 1$ ....②

To eliminate y, multiply ② by 2
$6x - 2y = 2$ ....③
Add equations ① and ③
$7x = 7 \Rightarrow x = 1$
Substitute in equation ②
$3 - y = 1 \Rightarrow y = 2$

(b) $2x + 3y = 7$ ....①
$5x - 2y = 8$ ....②
Since the LCM of 2 and 3 is 6, multiply ① by 2 and ② by 3
$4x + 6y = 14$ ....③
$15x - 6y = 24$ ....④
③ + ④: $19x = 38 \Rightarrow x = 2$
Substitute in equation ①
$2(2) + 3y = 7$
$4 + 3y = 7$
$3y = 3 \Rightarrow y = 1$

Module 14, Topic 11: Equations

Method of Substitution

 Example

Solve the simultaneous equations in the examples above using the method of substitution.

Solutions

(a) $x + 2y = 5$ .... ①
     $3x - y = 1$ .... ②
     Make x the subject in equations ①
         $x = 5 - 2y$ .... ③
     Substitute in equation ②
     $3(5 - 2y) - y = 1$
     $15 - 6y - y = 1$
           $-7y = -14 \Rightarrow y = 2$
     Substitute in equation ③
         $x = 5 - 2(2) \Rightarrow x = 1$

(b) $2x + 3y = 7$ .... ①
     $5x - 2y = 8$ .... ②
     From equation ①, $2x = 7 - 3y \Rightarrow x = \dfrac{7 - 3y}{2}$ .... ③
     Substitute in equation ②
         $5\left(\dfrac{7 - 3y}{2}\right) - 2y = 8$
         $5(7 - 3y) - 4y = 16$
         $35 - 15y - 4y = 16$
             $-19y = -19 \Rightarrow y = 1$
     Substitute in equation ③
         $x = \dfrac{7 - 3(1)}{2} \Rightarrow x = 2.$

Exercise 11:3

1. Solve the following simultaneous equations using the method of elimination.

(a) $p - 3q = 10$ $3p - 2q = 16$	(b) $s = 2t - 1$ $2s = 3t + 2$	(c) $5x - 2y = 14$ $2x + 2y = 14$
(d) $4m + 4n = 3$ $m + 2n = 1$	(e) $2x + y = 7$ $3x - 2y = 7$	(f) $3u - 7v = 1$ $2u + v = 12$

2. Solve the simultaneous equations in problem 1, using the method of substitution.
3. Determine which method is better and use it to solve each of the following simultaneous equations.

(a) $3a + 5b = 4$
    $2a + 3b = 4$

(b) $x + y = 1$
    $x - y = 3$

(c) $3x + 4y = 0$
    $x = 2y - 5$

(d) $y = x + 1$
    $x + y = 3$

(e) $2x + y = 7$
    $3x - 2y = 7$

(f) $3u - 7v = 1$
    $2u + v = 12$

(g) $x + 2y = 8$
    $x - 3y = 3$

(h) $x - 2y = 0$
    $x + 3y = -10$

(i) $4x = y + 7$
    $3x + 4y + 9 = 0$

(j) $x + y = 7$
    $x - y = 1$

(k) $x + y = 4$
    $-x + y = 2$

(l) $5a + 3b = 12$
    $a - 3b = 6$

(m) $x + y = 7$
    $x - y = 3$

(n) $x - y = 3$
    $2x + y = 12$

(o) $5a + 3b = 1$
    $2a + 3b = -5$

(p) $x + y = 11$
    $x - y = 5$

(q) $x + y = -2$
    $x - y = 0$

(r) $3r + 5s = 21$
    $7r - 2s = 8$

11.5 Simultaneous Equations with Fractions and Decimals

Method of Elimination

In solving simultaneous equations involving fractions and decimals, first get rid of the fractions by multiplying each term by the LCM of the denominators (or a required power of 10 in the case of decimals). After this, use the method of elimination or substitution.

Module 14, Topic 11: Equations

 Example

Solve the following simultaneous equations:

(a) $\dfrac{x}{3}+\dfrac{y}{4}=\dfrac{1}{12}$

$\dfrac{3x}{2}-\dfrac{y}{3}=-4$

(b) $0.03x+0.05y=51$

$0.8x-0.7y=140$

Solutions

To solve these equations by either method, it is advisable first to eliminate the fractions or decimals.

(a) $\dfrac{x}{3}+\dfrac{y}{4}=\dfrac{1}{12}$ ....①

$\dfrac{3x}{2}-\dfrac{y}{3}=-4$ ....②

Since the LCM of 3, 4 and 12 is 12 and the LCM of 2 and 3 is 6, multiply equation ① by 12 and equation ② by 6:

$4x+3y=1$ ....③

$9x-2y=-24$ ....④

Now solve the simultaneous equations using any of the methods desired.

By Method of Elimination

Multiply equation ③ by 2 and equation ④ by 3.

$8x+6y=2$ ....⑤

$27x-6y=-72$ ....⑥

⑥−⑤: $35x=-70 \Rightarrow x=-2$

Substitute in ③

$4(-2)+3y=1$

$3y=9 \Rightarrow y=3$

By Method of Substitution

From ③, $y=\dfrac{1-4x}{3}$ ....⑦

Substitute in ④: $9x-2\left(\dfrac{1-4x}{3}\right)=-24$

Multiply both sides by 3

$$27x - 2(1-4x) = -72$$
$$27x - 2 + 8x = -72$$
$$35x = -70 \Rightarrow x = -2$$

Substitute in equation ⑦

$$y = \frac{1 - 4(-2)}{3} \Rightarrow y = 3$$

(b) $0.03x + 0.05y = 51$ ....①
$0.8x - 0.7y = 140$ ....②

Multiply equation ① by 100 and equation ② by 10.

$$3x + 5y = 5100 ....③$$
$$8x - 7y = 1400 ....④$$

We can then solve the equations ③ and ④ using any of the two methods as follows.

By Method of Elimination

Multiply equation ③ by 7 and equation ④ by 5

$$21x + 35y = 35700 ....⑤$$
$$40x - 35y = 7000 ....⑥$$

Add equation ⑥ and equation ⑤

$$61x = 42700 \Rightarrow x = 700$$

Substitute in ③:

$$3(700) + 5y = 5100$$
$$2100 + 5y = 5100$$
$$5y = 3000 \Rightarrow y = 600$$

By Method of Substitution

From ③, $5y = 5100 - 3x$

$$y = \frac{5100 - 3x}{5} ....⑦$$

Substitute in equation ④: $8x - 7\left(\dfrac{5100 - 3x}{5}\right) = 1400$

$$40x - 7(5100 - 3x) = 7000$$
$$40x - 35700 + 21x = 7000$$
$$61x = 42700 \Rightarrow x = 700$$

Substitute in equation ⑤: $y = \dfrac{5100 - 3(700)}{5} \Rightarrow y = 600$

 Exercise 11:4

Solve the following simultaneous equations.

1. $\dfrac{x}{3} - 2y = 1$
   $x + 3y = 12$

2. $\dfrac{x}{3} + \dfrac{y}{4} = \dfrac{1}{12}$
   $2x - 3y = -4$

3. $3a - b = 500$
   $0.7a + 0.2b = 550$

4. $x - 2y = 500$
   $0.03x + 0.02y = 51$

5. $m = 4n - 100$
   $0.06m = 0.05n + 32$

6. $0.03x + 0.04y = 44$
   $0.04x + 0.02y = 42$

7. $0.8p - 0.7q = 140$
   $0.03p + 0.05q = 51$

8. $0.05(u + 2000) = 0.3(v + 3000)$
   $u = \dfrac{v}{2} + 500$

11.6 Simultaneous Linear Equations in Real Life

We are now armed to solve real life problems such as the needle and razor problem at the beginning of this section. First we have to translate the problem into a pair of simultaneous linear equations as follows.

Let r represent the cost of a razor and n the cost of a needle. Then,

$2r + 3n = 70$ .... ①
$3r + 5n = 110$ .... ②

From equation ①: $r = \dfrac{70 - 3n}{2}$ .... ③

Substitute equation ③ in equation ②,

$$3\left(\frac{70-3n}{2}\right) + 5n = 110$$

Multipling bothsides by 2 and opening the bracket,
$$210 - 9n + 10n = 220$$
$$210 + n = 220$$
$$210 + n = 220$$
$$\Rightarrow n = 10 \text{ Francs.}$$

Substitute equation ③: $r = \frac{70-3(10)}{2} \Rightarrow r = 20$ Francs.

Therefore, 4 razors and 7 needles $= 4r + 7n$
$$= 4(20) + 7(10) = 150 \text{ Francs.}$$

Example

1. A credit union gave Mr. Ngong 16000 francs consisting of 500 francs coins and 100 francs coins. The number of 100 francs coins is three times the number of 500 francs coins. How many of each type of coin did Mr. Ngong receive?

   Solution
   Let h = number of 100 francs coins and
   f = number of 500 francs coins
   Then $100h + 500f = 16000$
   $$\Rightarrow h + 5f = 160 \quad \text{...........................①}$$
   Also $h = 3f$ ....②
   Substitute ② in ①:
   $$\Rightarrow 3f + 5f = 160$$
   $$8f = 160 \Rightarrow f = 20$$
   Substitute in ②: $h = 3(20) = 60$
   Number of 100 francs coins = 60
   Number of 500 francs coins = 20

2. A father gave his twin children 350 francs each. One bought 3 packets of biscuits, 2 sweets, and the other bought 2 packets of biscuits and 6 sweets.
   (b) Find the cost of a packet of biscuits. Find the cost of a single sweet.

   Solution
   $$3b + 2s = 350 \quad \text{.......................①}$$
   $$2b + 6s = 350 \quad \text{.......................②}$$
   From ②, $b + 3s = 175$
   $$\Rightarrow b = 175 - 3s \quad \text{.....................③}$$
   Substitute in ①: $3(175 - 3s) + 2s = 350$
   $$\Rightarrow 525 - 9s + 2s = 350$$
   $$525 - 7s = 350$$

$-7s = -175 \Rightarrow s = 25$ francs

Substitute in ③:
$b = 175 - 3(25) = 100$ francs

(a) a packet of biscuits = 100 francs   (b) a single sweet = 25 francs.

3. Two cars leave Makenene at the same time travelling in opposite directions. One is travelling at 80 km/h and the other is travelling at 70 km/h. How long will it take the two cars to be 300 km apart?

Solution

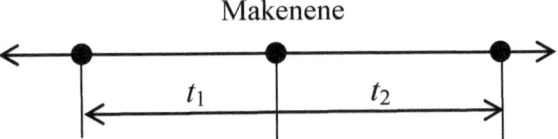

Let the times taken by the two cars be t_1 and t_2.

Then $t_1 = t_2$ ....①

Since distance = speed × time

$80t_1 + 70t_2 = 300 \Rightarrow 8t_1 + 7t_2 = 30$ ....②

Substitute ① in ②: $8t_1 + 7t_1 = 30$

$15t_1 = 30 \Rightarrow t_1 = 2$

Therefore, it will take 2 hours for the two cars to be 300 km apart.

 Exercise 11:5

1. A student bought 3 books and 2 pencils for 340 FRS. Another student bought 2 books and 3 pencils of same kind for 260 FRS. Find the cost of a book and a pencil.
2. The value of the expression $mx^2 + nx$ is 8 when $x = 2$, and 27 when $x = 3$. Determine the value of n.
3. The sum of two numbers is 26 and their difference is 28. Find them.
4. The average of two numbers is 13. The difference of the two numbers is 6. Find the two numbers.
5. A bottle and a cork together weigh 18 g. Given that, the bottle weighs 16 g more than the cork, find the weight of each.
6. 3 nuts and 4 bolts have a mass of 72 g. 4 nuts and 5 bolts have a mass of 94g. Find the mass of   (a) a nut   (b) a bolt
7. Three coconuts and two oranges cost 430 FRS. One coconut and four oranges cost 210 FRS. Find the cost of each coconut and each orange.
8. A man is 9 times as old as his son is. In four years time, he would be 5 times as old as his son would. What is the present age of both of them in years?
9. A hotel has first-class rooms, which cost 4500 francs, and second-class rooms,

which cost 2500 francs. On a certain day, the manager gave out 32 rooms at a total cost of 130,000 francs.
  (a) How many first-class rooms and second-class rooms did the manager give out on that day?
  (b) Calculate the total amount collected for each class of rooms.
10. A tailor takes 8 hours to stitch a trouser, which cost 4800 francs. The tailor takes 6 hours to stitch a shirt, which cost 4000 francs. The tailor used 120 hours to stitch shirts and trousers, which he sold for 78400 francs. How many shirts and how many trousers did he stitch?
11. A man left his house jogging at 8 km/h. On his return, he was tired and could only run at 5 km/h. The total time used for this exercise was 1 and a half hours. How far did he go from his house?

11.7 Quadratic Equations

 Group Activity

The cost of a carton of macaroni is normally 4000 Francs. Due to a 20 Francs discount per sachet, a woman buys 10 more sachets at the same amount. You need to organize a party which will be attended by 240 people and you estimate that one person will eat half a sachet of macaroni. How much will you budget for this item and how many sachets of macaroni will you expect to be bought?

This is another very simple practical real life problem that occurs almost on a daily bases. People who have studied quadratic equations will very easily tackle these types of problems without any ado. However if you don't know how to solve quadratic equations you may not solve this easy problem.

A quadratic equations is an equations in which the highest power of the unknown is 2.

An example of a quadratic equations is $2x^2 + 5x + 2 = 0$

Before attempting the macaroni problem, lets first study how to solve quadratic equations.

11.8 Standard Form Quadratic Equations

A quadratic equation is any second-degree polynomial equation (i.e. any polynomial equation where the highest power of the unknown is 2). The standard form of a quadratic equation is $ax^2 + bx + c = 0$, where $a \neq 0$, b and c are constants. There are four basic methods of solving Standard form quadratic equations. These include the factorization method, the method of completing the square, the formula method and the graphical method. This topic examines the first three methods and Topic 34 examines the graphical method.

11.9 Factorization method

If the expression $ax^2 + bx + c$ is factorable, the result will be of the form $(px + q)(rx + s) = 0$.

The fundamental theorem for solving the equation $(px + q)(rx + s) = 0$ is that either $(px + q) = 0$ or $(rx + s) = 0$

$$\Rightarrow x = -\frac{q}{p} \quad \text{or} \quad x = -\frac{s}{r}$$

Example

1. Find the value of x if $(x - 1)(x + 2) = 0$.

   Solution
   $$(x - 1)(x + 2) = 0$$
   Either $x - 1 = 0 \Rightarrow x = 1$
   or $x + 2 = 0 \Rightarrow x = -2$.

2. Solve the following quadratic equations using the factorization method.
   (a) $6x^2 + 7x - 3 = 0$ \qquad (b) $x^2 + 9x + 18 = 0$

   Solution

   (a)
   $$6x^2 + 7x - 3 = 0$$
   $$6x^2 + 9x - 2x - 3 = 0$$
   $$3x(2x + 3) - 1(2x + 3) = 0$$
   $$(3x - 1)(2x + 3) = 0$$
   $\therefore (3x - 1) = 0$ or $(2x + 3) = 0$
   $$x = -\frac{3}{2} \text{ or } x = \frac{1}{3}$$

   (b)
   $$x^2 + 9x + 18 = 0$$
   $$(x + 3)(x + 6) = 0$$
   $\therefore (x + 3) = 0$ or $(x + 6) = 0$
   $$x = -6 \text{ or } x = -3$$

Exercise 11:6

Solve the following quadratic equations using the factorization method.
(1) $x^2 - 2x - 8 = 0$ (2) $x^2 + x - 2 = 0$ (3) $x^2 - 5x + 6 = 0$
(4) $2x^2 - x - 1 = 0$ (5) $3x^2 - 12x - 63 = 0$ (6) $6x^2 - 5x - 6 = 0$
(7) $12x^2 + 18x - 15 = 0$ (8) $2x^2 - 9x - 5 = 0$

11.10 The Quadratic Formula

The following formula, called the **quadratic formula** can also be used to find the roots of any quadratic equation especially when the quadratic expression is not factorable.

The **quadratic formula** states that if $ax^2 + bx + c = 0$, where a, b and c are constants and $a \neq 0$, then $x = \dfrac{-b \pm \sqrt{b^2 - 4ac}}{2a}$.

Take note that the division bar goes across!

i.e. $x \neq -b \pm \dfrac{\sqrt{b^2 - 4ac}}{2a}$

 Example

Use the quadratic formula to solve the equation $2x^2 + 3x - 2 = 0$.

Solution

$$x = \frac{-b \pm \sqrt{b^2 - 4ac}}{2a}, a = 2, b = 3, c = -2$$

$$x = \frac{-3 \pm \sqrt{3^2 - 4(2)(-2)}}{2(2)} = \frac{-3 \pm \sqrt{25}}{4}$$

$$x = \frac{1}{2} \text{ or } x = -2$$

Module 14, Topic 11: Equations

 Exercise 11:7

Solve the following quadratic equations using the formula method, leaving your answer to 1 decimal places.
1. $2x^2 - 5x - 4 = 0$ 2. $x^2 + 7x + 5 = 0$ 3. $2x^2 + 5x + 1 = 0$ 4. $x^2 + 6x - 10 = 0$ 5. $3x^2 - 4x - 2 = 0$ 6. $6x^2 - 10x + 3 = 0$ 7. $5x^2 - 10x + 4 = 0$ 8. $9x^2 + x - 2 = 0$

11.11 Problems leading to Quadratic Equations

Many real life problems such as the macaroni problem at the beginning of this section involve quadratic equations and until the solving begins, it will not be easy to predict. Before continuing, let's now solve the macaroni problem.

Let x be the number of sachets per carton.

Then cost of each sachet $= \dfrac{4000}{x}$ Francs.

If the woman buys 10 more sachets at the same amount (4000 FRS),

Then cost of each discounted sachet $= \dfrac{4000}{x+10}$ Francs.

But cost per discounted sachet is 20 Francs less.

$\therefore \quad \dfrac{4000}{x} - \dfrac{4000}{x+10} = 20$

Multiply both sides by the LCM (i.e. by $x(x+10)$)

$4000(x + 10) - 4000x = 20x(x + 10)$
$4000x + 40000 - 4000x = 20x^2 + 200x$
$40000 = 20x^2 + 200x$

Dividing both sides by 20 and rearranging leads to $x^2 + 10x - 2000 = 0$, which we can solve using any of the methods we have learnt.

$$(x + 50)(x - 40) = 0$$
$$x = -50 \text{ or } x = 40$$

The number of sachets per carton cannot be negative, so $x = 40$ and $x \neq -50$.

Then cost of each discounted sachet $= \dfrac{4000}{x+10} = \dfrac{4000}{40+10} = 80$ Francs.

Number of sachets required by 240 people $= \dfrac{1}{2} \times 240 = 120$ sachets

Therefore amount to be spent on this item $= 120 \times 80 = 9600$ Francs.

Competency Based Mathematics for Secondary Schools. Book 3

 Example

A man wants to buy a piece of land. The owner tells him that the area of the piece of land is 80 square metres and that the length is 2 metres longer than the width. The man hires you to work out the length and width for him. Go ahead and do your job.

Solution
Let the width be x m, then the length will be $(x + 2)$ m.

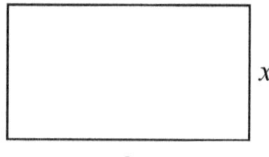

$x(x+2) = 80 \Leftrightarrow x^2 + 2x = 80$
$\Leftrightarrow x^2 + 2x - 80 = 0$ or $(x-8)(x+10) = 0$
$\Leftrightarrow x = 8$ or $x = -10$

Since $x > 0$, so the solution $x = -10$ should be discarded. Clearly if the width is 8 m, the length will be 10 m, since $8 \times 10 = 80$.

 Exercise 11:8

Translate and solve the following equations.
1. The sum of the squares of two consecutive integers is 13. Find the numbers.
2. The square of a number is 12 more than the number. Find the number and its square.
3. At the beginning of a school year, a man bought for his two children $(x - 5)$ exercise books at $(3x + 75)$ FCFA each. Given that, he spent 2025 FCFA altogether, find the number of exercise books he bought and the exact price of each.
4. A woman bought a piece of land in Bamenda. While in Yaoundé, she decided to sell it. The buyer was interested in knowing the dimensions of the piece of land. What the woman could remember was that its area is 247.5 square metres and the length is 1.5 m more than the width. Help the woman out of this problem.
5. The formula $T_n = \dfrac{n(n+1)}{2}$ is a formula for finding any triangular number. Find the value of n for which $T_n = 78$.
6. The product of two positive integers is 27. One is three more than twice the other is. Find the numbers.
7. The capacity of a hall is 144 persons. The number of rows is 7 more the number

Module 14, Topic 11: Equations

of seats in each row. Determine the number of rows of seats in the hall.
8. The area and perimeter of a rectangular farm are 24 square metres and 20 metres respectively. Find the length and width of the farm.
9. The height of a triangular flowerbed is 4 m less than the base. The area of the flowerbed is 48 m². Calculate the height and the length of the base of the flowerbed.
10. The product of a positive integer and a number three less than the integer is equal to the integer increased by 32. What is the integer?
11. Two positive integers differ by 8. The sum of the square of the larger and the smaller is 124. Find the numbers.
12. During a procession, the signboard bearer and the choir leader are alone on their own rows. The flag bearer and shield bearer occupy the next row. The arrangement of the rest of the choir members is such that the number of choristers on each row exceeds the number of rows by three. There are altogether 112 choristers. Find the total number of rows on this procession.

Multiple Choice Exercise 11

1. The pair of values of x and y which satisfy the simultaneous equations $x + y = 3$ and $3x - y = 1$ are:
   [A] (1, –2)   [B] (2, –1)   [C] (1, 2)   [D] (–2, 1)

2. Given that $x = 2y - 1$ and $2x = 2y - 1$. The value of x is:
   [A] –9   [B] –4   [C] 9   [D] 0

3. If $x = 2y - 1$ and $2x = 3y + 2$, then y equals:
   [A] 4   [B] 9   [C] –4   [D] –9

4. The roots of the simultaneous equations $x + y = 4$ and $2y - x = 5$ are:
   [A] (–1, 3)   [B] (–1, –3)   [C] (1, –3)   [D] (1, 3)

5. The values of x and y that satisfy the simultaneous equations $2x + y = 7$ and $3x - 2y = 7$ are:
   [A] (–1, 3)   [B] (3, 1)   [C] (–1, –3)   [D] (3, –1)

6. Given that $2p - m = 6$ and $2p + m = 1$. The value of $4p + 3m$ is:
   [A] 1   [B] 3   [C] 5   [D] 7

7. If $x + y = \dfrac{3}{2}$ and $x - y = \dfrac{5}{2}$, then $2y + x$ equals:
   [A] –2   [B] 1   [C] $\dfrac{1}{2}$ [D] –1

8. If $x + 2y = 1$ and $x - y = 2$, the value of $x + y$ is:
   [A] $1\dfrac{1}{3}$ [B] 1   [C] –1   [D] $-1\dfrac{1}{3}$

9. Given that $x + y = 7$ and $3x - y = 5$. When evaluated $\frac{y}{2} - 3$ gives:
   [A] 3 [B] 1 [C] −1 [D] 4

10. If $2x + y = 7$ and $3x - y = 3$, then $7x$ is greater than 10 by:
    [A] 1 [B] 7 [C] 3 [D] 10

11. Given the equations $4y - 5x = 14$ and $y = 3x$. The values of x and y are respectively:
    [A] (2, 6) [B] (−2, −6) [C] (2, −6) [D] (−2, 6)

12. The values of x and y which satisfy the simultaneous equations $4x - y = 11$ and $5x + 2y = 4$ are:
    [A] $x = 2, y = 3$ [B] $x = -2, y = -3$
    [C] $x = -2, y = 3$ [D] $x = 2, y = -3$

13. If $3p - q = 6$ and $2p + q = 4$, then q is equal to:
    [A] 0 [B] $\frac{1}{2}$ [C] $\frac{2}{3}$ [D] 1

14. The values of $x - y$, which satisfy the simultaneous equations $4x - 3y = 7$ and $3x - 2y = 5$ are:
    [A] −3 [B] 3 [C] 2 [D] −2

15. $\frac{1}{3}x + y = 3$ and $x + \frac{1}{2}y = 4$ provided:
    [A] $x = 3, y = 3$ [B] $x = 3, y = -3$
    [C] $x = -3, y = 3$ [D] $x = -3, y = -3$

16. If the solutions of the pair of equations $2x + 3y = p$ and $3x - y = q$ are $x = -1$ and $y = 2$, the values of p and q must be:
    [A] $p = -4, q = 5$ [B] $p = -5, q = 4$
    [C] $p = -4, q = -5$ [D] $p = 4, q = -5$

17. An exercise book and a pencil cost 180 F. If the exercise book costs 140 F more than the pencil, then the pencil costs:
    [A] 30 F [B] 25 F [C] 20 F [D] 15 F

18. Ambe is four times as old as Ndeh. If the sum of their ages is 20 years, the difference in their ages in years is:
    [A] 16 [B] 12 [C] 8 [D] 4

19. 2 nuts and 3 bolts have a mass 28 g. 3 nuts and a bolt have a mass of 21 g. The mass of a bolt is:
    [A] 4 g [B] 5 g [C] 6 g [D] 7 g

20. The function, which is not a quadratic in x is:
    [A] $y = 2x^2 - 5x$ [B] $y = x(x - 5)$ [C] $y = x^2 - 5$ [D] $y = 5(x - 1)$

21. If $x^2 + 15x + 50 = ax^2 + bx + c = 0$, it is not true to say that:
    [A] $x = 5$ [B] $x = 10$ [C] $x + 10 = 0$ [D] $bc = 750$

22. If $2x^2 + kx - 14 = (x + 2)(2x - 7)$ then, the value of k is:
    [A] −3 [B] 5 [C] 9 [D] 11

23. Given that $5x^2 + 4x + 3 = a + b(x+1) + cx(x+1)$. The value of the constants a, b, and c are respectively:
    [A] 7, 4, 5   [B] 4, −1, 5   [C] −1, 5, 4   [D] 7, 5, 4
24. The equation $(x+2)(x-7) = 0$ has roots:
    [A] −2 and 7   [B] 2 and −7   [C] −2 and −7   [D] 2 and 7
25. If the roots of the equation $(3x - 1)(x + 2) = 0$ are p and q, then, the value of $p + q$ is:
    [A] $2\frac{1}{2}$ [B] $1\frac{2}{3}$ [C] $-1\frac{2}{3}$ [D] $-2\frac{1}{2}$
26. The values of x, which satisfy, $x^2 + 2x + 1 = 25$ are:
    [A] −6, −4   [B] 6, −4   [C] 6, 4   [D] −6, 4
27. The values of x, which satisfy the equation, $x^2 - 2x - 3 = 0$ are:
    [A] (−3, 1)   [B] (−1, 3)   [C] (−3, 1)   D] (−1, −3)
28. The smaller value of x for which $x^2 - 3x + 2 = 0$ is:
    [A] 1   [B] 2   [C] −1   [D] −2
29. $6x^2 - 7x - 5 = 0$, only if:
    [A] $x = \frac{1}{2}, -2\frac{1}{2}$ [B] $x = \frac{1}{3}, 2\frac{1}{2}$
    [C] $x = 1\frac{2}{3}, -\frac{1}{2}$ [D] $x = -1\frac{2}{3}, \frac{1}{2}$
30. $2x^2 - 3x - 2 = 0$ is true if and only if:
    [A] $x = -2, \frac{1}{2}$ [B] $x = 1, 8$ [C] $x = -\frac{1}{2}, 2$ [D] $x = -1, 2$
31. $2a^2 - 3a - 27 = 0 \Leftrightarrow a =$:
    [A] $x = -3, -\frac{9}{2}$ [B] $x = -\frac{2}{3}, 9$ [C] $x = 3, \frac{9}{2}$ [D] $x = -3, -\frac{9}{2}$
32. The equation $3x^2 + 25x - 18 = 0$ has roots:
    [A] −3, 2   [B] $-9, \frac{2}{3}$ [C] $-\frac{3}{2}, 9$ [D] −2, 3
33. The sum of the roots of the equation $2x^2 + 3x - 9 = 0$ is:
    [A] −18   [B] −6   [C] $\frac{9}{2}$ [D] $-\frac{3}{2}$
34. The two values of x that satisfy the equation $5x^2 - 4x - 1 = 0$ are:
    [A] $1, -\frac{1}{5}$ [B] $-1, -\frac{1}{5}$ [C] $-1, \frac{1}{5}$ [D] $1, \frac{1}{5}$
35. The solutions of the equation $3 + 5x - 2x^2 = 0$ are:
    [A] $-\frac{1}{2}, -3$ [B] 2, 3   [C] −2, 3   [D] $-\frac{1}{2}, 3$
36. Given that $10 - 3x - x^2 = 0$. The values of x are:
    [A] $x = 2$ or −5   [B] $x = -2$ or 5   [C] $x = -1$ or 10   [D] $x = 2$ or 5
37. The equation $3a + 10 = a^2$ gives rise to the roots:
    [A] $a = 5$ or 2   [B] $a = -5$ or 2   [C] $a = -5$ or −2   [D] $a = 5$ or −2

38. One of the roots of the equation $6x^2 = 5 - 7x$ is:
    [A] $-\dfrac{1}{2}$ [B] $-\dfrac{1}{3}$ [C] $\dfrac{1}{2}$ [D] $2\dfrac{1}{2}$

39. The values of y, which satisfy the equation $3y^2 = 3y$ are:
    [A] $y = -3$ or 9 [B] $y = 0$ or 9 [C] $y = -3$ or 3 [D] $y = 3$ or 9

40. The equation $7y^2 = 27y$ has roots:
    [A] $y = 3$ and $y = 7$ [B] $y = 0$ and $y = 7$
    [C] $y = 0$ and $y = \dfrac{3}{7}$ [D] $y = 0$ and $y = 9$

41. A root of the equation $x^2 + 6x = 0$ is:
    [A] 0 [B] 6 [C] 2 [D] 3

42. The value of k, which makes the expression $m^2 - 8m + k$ a perfect square, is:
    [A] 2 [B] 4 [C] 8 [D] 16

43. For $x^2 - 6x$ to be a perfect square:
    [A] add 9 [B] add 36 [C] 1 [D] add 3

44. The number to the expression $x^2 - 8x$ to make a perfect square is:
    [A] 36 [B] 9 [C] 25 [D] 16

45. On subtracting five times a certain integer from twice the square of the integer, the result is 63. The integer is:
    [A] 21 [B] 7 [C] 9 [D] 4

46. The area of a rectangle is the product of its length and breadth. The length and breadth of a rectangle are $(x-3)$ cm and $(x-5)$ cm. The area of the rectangle is 24 cm² only if:
    [A] $x = -9$ or -1
    [B] $x = 9$ or 1
    [C] $x = 9$ or -1
    [D] $x = -9$ or 1

47. The sum of the squares of two natural numbers is 29. One of the numbers is three more than the other is. The larger of the numbers is:
    [A] -5 [B] -2 [C] 2 [D] 5

48. 120 soldiers are standing in rows. There are 2 more soldiers in each row than there are rows. The number of rows is:
    [A] 8 [B] 10 [C] 12 [D] 55

49. The difference between two numbers is 6 and the difference between their squares is 132. The numbers are:
    [A] 8, 14 [B] $-8, -14$ [C] $-8, 14$ [D] $8, -14$

50. A man uses 50 m of fencing to fence a garden at his back yard with his house as one side of the fence. If the area of the garden is 300 square metres, the length of the garden is:
    [A] 6 [B] 10 [C] 15 [D] 25

Topic 12

LOGIC

Objectives
At the end of this topic, the learner should be able to:

1. Draw up truth tables.
2. Use logic connectives;
3. State and prove De Morgan's laws using truth tables.
4. Use truth tables to show that two statements are logically equivalent.

12.1 The Concept of Logic

Logic is a science, which deals with the principles of valid reasoning and argument. Logic is very important in disciplines such as mathematics, law, computer science etc.

12.2 Statements or Propositions

A **statement** or **proposition** is a sentence that is either true or false but not both. This definition lays emphasis on the fact that a statement must contain enough information for anyone to decide whether it is true or false. The statement itself, the context, the time or the place may provide this information. Questions, commands and remarks are therefore not statements because one cannot decide whether they are true or false. We usually denote statements by lower case letters $a, b, c, \cdots$

 Example

State giving reasons whether or not, the following are statements.
p: Bamenda is not in Cameroon.
q: In which country is Bamenda found?
r: Let us eat the rice!
s: That is exceptional of you!
t: Time for breakfast! Bring tea and bread!

Solution
Among the above, the only statement, which though is false, is "p".
"q" is not a statement. It is a question.
"r" is not a statement. It is a command.
"s" is not a statement. It is a remark. Its truth depends on the judge and the addressee.
"t" is not a statement, It is a reminder and a command.

Questions, commands, remarks and reminders are not statements because they are neither true nor false.

12.3 The Truth Value of a Statement

The **truth-value** of a statement is the truthfulness or falsity of the statement. We denote the truth-value of a true statement by T or 1, and denote the truth-value of a false statement by F or 0. A truth table is a table, which shows the truth-values of a statement. Thus for a statement p, the truth table could be as follows.

Module 14, Topic 12: Logic

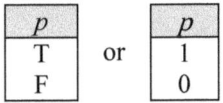

Established statements, which have been established by means of a proof, are called **theorems** and their truth-value is T or 1.
The following are some examples of theorems.

t: The sum of the interior angles of a triangle is $180°$.
s: The exterior angle of a triangle is equal to the sum of the two interior opposite angles.

12.4 Closed and Open Statements

A closed statement is a statement concerning a definite object. An open statement on the other hand is a statement that does not concern a definite object.

Consider the following statements.
a: Mr. Tantoh has four Children.
b: He is a teacher.
c: $4 + 5 = 9$
d: $2 \times 3 = 8$
e: $3x - 1 = 11$
f: $4p \geq 1 = 11$

Statements c, d, e and f are examples of mathematical statements. Statements a, c and d are closed while statements b, e and f are open. The "he" in statement b may stand for any member in a group of male teachers. We can close this statement by replacing the "he" by the name of any male teacher in the reference group. The statement will then be either true or false. We can substitute the x and p in statements e and f with numbers to close them and render them true or false.

12.5 Domain and Variable

A **variable** is the open part, which we must close in an open statement to make it true or false. The **domain** or the **replacement set** of the variable is the set of all the possible values, which the variable can take.
For instance;
If $x = 4$, statement e is true, otherwise it is false.
If $p \geq 3$, $p \in \mathbb{R}$, statement f will be true, otherwise it is false. The **truth** or **solution set** of a statement is the set of values of a variable, which makes the statement true.

Thus, the truth set of statements *e* and *f* are $\{x = 4\}$ and $\{p \geq 3, p \in \mathbb{R}\}$ respectively.

 Example

State the domain and the truth set for each of the following statements.
(a) $5x + 2 = 13$ (b) $7y < 42$ (c) $(x + 1)(x - 2) = 0$

Solution
(a) The domain is $\mathbb{Q}$, the set of rational numbers and the truth set is $x : x = \frac{11}{5}, x \in \mathbb{Q}$.
(b) The domain is $\mathbb{R}$, the set of natural numbers and the truth set is $\{y : y < 6, y \in \mathbb{R}\}$.
(c) The domain is $\mathbb{Z}$, the set of integers and the truth set is $\{-1, 2\}$.

 Exercise 12:1

1. Which of the following is a true, false or open statement?
   (a) $3 + 3 = 3 \times 2$
   (b) $8 \times 1 \neq 8 + 1$
   (c) $21 + 0 = 21 \times 1$
   (d) $3 + 7 = 7 + 3$
   (e) The sum of 25 and 0 is 25.
   (f) The sum of a whole number and 0 is the number.
   (g) The sum of a whole number and 7 is 7.
   (h) The product of 67 and 1 is 68.
   (i) If 3 is subtracted from 20, the difference is the same as when 20 is subtracted from 3.
   (j) The product of 8 and 4 is the same as the product of 4 and 8.
   (k) If 8, is divided by 4, the quotient is the same as when 4 is divided by 8.
2. Given that the domain of the variable x is $A = \{1, 2, 3, 4, 5, 6, 7, 8, 9\}$. State the truth or solution set of each of the following statements.
   (a) $x + 3 = 10$
   (b) $x > 7$
   (c) x is an even number
   (d) $x^2 = 25$
   (e) $x + 10 = 10 + x$
   (f) $x - 8 = 8 - x$
   (g) $0 \times x = x$
   (h) $0 \times x = 0$
   (i) $2x = x + 5$
   (j) $2x \neq x + 5$
   (k) $\{x : x \geq 7\}$
   (l) $\{x : x > 4\}$
3. Sort out the sentences, which are statements and state whether they are true or false.
   (a) Give me that pencil.
   (b) There is an Ocean in Bamenda.
   (c) All teachers are lazy.
   (d) All Bamenda people eat Achu.
   (e) Make sure your uniforms are neat.
   (f) Is Nigeria an African Country?
   (g) History is a science subject.
   (h) Ahidjo was a president of Cameroon.
4. Classify the following statements as closed or open statements.

Module 14, Topic 12: Logic

  (a) Mr. Fonche died two years ago.   (b) 3+5 = 9   (c) They are lazy.
  (d) All Bamenda people eat Achu.   (e) 4 × 3 = 12
  (f) Nigeria is an African Country.   (g) History is a science subject.
  (h) He was the president of Cameroon.   (i) Everyone loves Mr. Paul Biya.
  (j) She comes from Nkambe.   (k) $2x - 1$

5. Given that $0 \leq x \leq 10$, $x \in \mathbb{N}$. State the truth set for the following statements.
   (a) $x > 5$.   (b) x is a factor of 10.   (c) x is an odd number.
   (d) x is an even number.   (e) x is a prime number   (f) x is a multiple of 3.

6. State the truth-value of each of the following statements.
   (a) 37 is a prime number.   (b) 105 is a multiple of 3.
   (c) All Bamenda people eat Achu.   (d) History is a science subject.
   (e) 4 × 3 = 12   (f) In the set of real numbers $a - b = b - a$
   (g) Nigeria is an African Country.

12.6 Negation, ~p

The negation of a statement p, denoted by $\sim p$ or $\neg p$ or p' and read 'not p' is the statement formed from p by inserting the word "not" into p or placing the phrases "it is false that"…or "it is not true that"…before the statement p.

 Example

Let p be the statement "Bamenda is in Cameroon". State the negation $\sim p$ of p.

Solution
$\sim p$: Bamenda is not in Cameroon or
$\sim p$: It is not true that Bamenda is in Cameroon or
$\sim p$: It is not true to say that Bamenda is in Cameroon or
$\sim p$: It is false that Bamenda is in Cameroon.

The fundamental property of a statement p and $\sim p$ is that,

If p is, true $\sim p$ is false; if p is false, $\sim p$ is true.

This means that p and $\sim p$ cannot be both true and both false. We can summarize this fundamental property on a truth table or Venn diagram as follows,

p	$\sim p$
T	F
F	T

or

p	$\sim p$
1	0
0	1

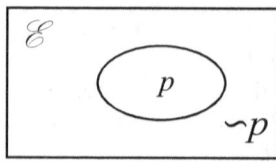

 Example

State the negation of the following statements.
q: It is raining.
r: All Cameroonians speak both English and French.
s: 5 is a multiple of 3.
t: $2x + 1 = 0$, for all values of x.

Solution
$\sim q$: It is not raining.
$\sim r$: It is not true that all Cameroonians speak both English and French.
$\sim s$: 5 is not a multiple of 3.
$\sim t$: It is not true that $2x + 1 = 0$, for all values of x.

 Exercise 12:2

1. State the negation of the following statements.
   (a) Mr. Fonche died two years ago.   (c) They are lazy.
   (d) All Bamenda people eat Achu.   (e) Loh cannot drive.
   (f) Nigeria is an African Country   (g) History is a science subject.
   (h) He was the president of Cameroon. (i) Everyone loves Mr. Paul Biya.
   (j) She comes from Nkambe.
   (k) Science has done more harm than good.
2. Write the negation of the following statements in symbolic language
   (a) $4y > 12$ (b) $3 + 5 = 9$ (c) $3p - 1 \geq 17$
   (d) $4 \times 3 = 12$ (e) $6x + 1 < 19$ (f) $2x - 1 = 0$
   (g) $A \not\subset B$ (h) $A \cap B \neq \emptyset$

12.7 Compound or Composite Statements

A compound or composite statement is a statement, which is made by combining two or more statements as in the following cases.

Module 14, Topic 12: Logic

12.8 Conjunction

When we make a composite statement by combining two statements with the use of the preposition "and" we call the resulting composite statement a **conjunction**. Symbolically, $p \wedge q$ denote a conjunction composed of two statements p and q.
We can this fundamental property on a truth table or Venn diagram as follows,

The fundamental property of the conjunction $p \wedge q$ is that $p \wedge q$ can only be true if both p and q are true otherwise the conjunction is false.

p	q	$p \wedge q$
T	T	T
T	F	F
F	T	F
F	F	F

or

p	q	$p \wedge q$
1	1	1
1	0	0
0	1	0
0	0	0

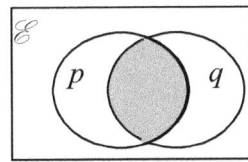

$p \wedge q$ is shaded

 Example

Given the statements
  x: Ngwa ate rice.
  y: Ngwa ate beans.
Make a conjunction, involving x and y.

Solution
  $x \wedge y$: Ngwa ate rice and beans.

We sometimes use the symbol $\wedge$ to define the intersection of sets. Thus,
$A \cap B = \{x : x \in A \wedge x \in B\}$.

12.9 Disjunction

When we make a composite statement by combining two statements with the use of the preposition "or" we call the resulting composite statement a **disjunction**.

Sometimes it is necessary to use the words 'either...or' for disjunctions. Symbolically, we denote a disjunction composed of the two statements p and q by $p \vee q$.

Consider the composite statement a below;

a: At 8 a.m. I shall be teaching in school or resting in my house.

Is it possible for someone to be teaching in school and resting in his house simultaneously? It is impossible!

Therefore, only one of the following simple statements b and c can be true of a.

b: At 8 a.m., I shall be teaching in school.
c: At 8 a.m., I shall be resting in my house.

An **exclusive disjunction** is a disjunction composed of two statements p and q, which cannot both be true.

The truth table and Venn diagram below represent exclusive disjunctions.

p	q	$p \vee q$
T	T	F
T	F	T
F	T	T
F	F	F

or

p	q	$p \vee q$
1	1	0
1	0	1
0	1	1
0	0	0

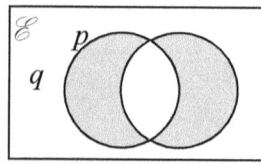

$p \vee q$ is shaded

Now consider the statement r below.

r: All the students who passed in English or French shall be given prizes.

Is it possible to pass in both English and French? Obviously! Therefore, any student who passes in both English and French should stand even a better chance to receive prizes. r equally carries the meaning in the statements s and t below.

s: All the students who passed in English shall be given prizes.
t: All the students who passed in French shall be given prizes.

Therefore, s is true, t is true and s and t are true.

An **inclusive disjunction** is a disjunction composed of two statements p and q, which can both be true. In logic a disjunction generally refers to inclusive disjunction where "or" is used in the sense of "and/or".

The fundamental property of the disjunction $p \vee q$ is that p is true or q is true or

both *p* and *q* are true otherwise, the conjunction is false.
We can summarize this fundamental property on a truth table or Venn diagram as follows.

p	*q*	*p* ∨ *q*
T	T	T
T	F	T
F	T	T
F	F	F

or

p	*q*	*p* ∨ *q*
1	1	1
1	0	1
0	1	1
0	0	0

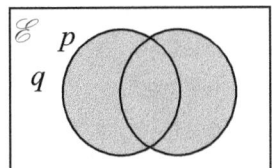

p ∨ *q* is shaded

We sometimes use the symbol ∨ to define the union of sets. Thus,
$A \cup B = \{x : x \in A \vee x \in B\}$.

12.10 Conditional Statements, $p \to q$

A **Conditional Statement** is a composite statement obtained by joining two statements *p* and *q* in such a way that if *p* is true *q* must be true. The conditional statement carrying the meaning "if *p* is true then *q* is true" is written symbolically as $p \to q$. We can read the conditional statement $p \to q$ in the following ways

(i) *p* implies *q*   (ii) *p* is sufficient for *q*
(iii) *p* only if *q*   (iv) *q* is necessary for *p*

If $p \to q$ is a theorem, we call the conditional statement an **implication** and use the symbol ⇒ instead of →.

> The fundamental property of a conditional statement
> $p \to q$ is that $p \to q$ is always true except *p* is true and *q* is false.

We can summarize this fundamental property on a truth table or Venn diagram as follows.

p	*q*	$p \to q$
T	T	T
T	F	F
F	T	T
F	F	F

or

p	*q*	$p \to q$
1	1	1
1	0	0
0	1	1
0	0	0

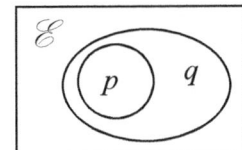

Example

Which of the following statements is false?
p: Cameroon is in Africa $\rightarrow$ $3 \times 2 = 7$.
q: Cameroon is in Europe $\rightarrow$ $3 \times 2 = 6$.
r: Cameroon is in Africa $\rightarrow$ $3 \times 2 = 6$.
s: Cameroon is in Europe $\rightarrow$ $3 \times 2 = 7$.

Solution
By the fundamental property of conditional statements, only p is false and q, r and s are all true.

12.11 Biconditional Statement, $p \Leftrightarrow q$ or $p \leftrightarrow q$

A **Biconditional Statement** is a proposition in logic involving two statements one of which is true if, and only if, the other is true.
The biconditional statement involving p and q is written symbolically as $p \Leftrightarrow q$ or $p \leftrightarrow q$. We can read the biconditional statement $p \Leftrightarrow q$ or $p \leftrightarrow q$ in the following ways
(i) p is a necessary and sufficient condition for q.
(ii) p implies and is implied by q
(iii) p if and only if q

We sometimes write the bi-conditional statement $p \Leftrightarrow q$ or $p \leftrightarrow q$ as p iff q.

> The fundamental property of a bi-conditional statement is that, either both statements are true or both are false, otherwise, the statement is false.

We can summarize this fundamental property on a truth table as follows.

p	q	$p \leftrightarrow q$
T	T	T
T	F	F
F	T	F
F	F	T

or

p	q	$p \leftrightarrow q$
1	1	1
1	0	0
0	1	0
0	0	1

 Example

Which of the following statements is true?
p: Cameroon is in Africa $\leftrightarrow$ $3 \times 2 = 7$.
q: Cameroon is in Europe $\leftrightarrow$ $3 \times 2 = 6$.

r: Cameroon is in Africa $\leftrightarrow 3 \times 2 = 6$.
s: Cameroon is in Europe $\leftrightarrow 3 \times 2 = 7$.

Solution
By the fundamental property of biconditional statements, r and s are true and p and q are false.

12.12 Logical Equivalence

Two statements p and q are **logically equivalent** if and only if they have the same truth tables. The statement p and q are logically equivalent is written symbolically as $p \equiv q$. The importance of truth tables cannot therefore be over emphasized. Apart from the use of truth tables to summarize composite statements, they are also used to determine if two statements are logically equivalent.

Investigative Activity

Draw truth tables showing the following:
(a) $p \wedge q$ (b) $p \vee q$ (c) $q \wedge p$ (d) $q \vee p$.

Hence, deduce that:
(i) $p \wedge q \equiv q \wedge p$ (ii) $p \vee q \equiv q \vee p$

Solution
(a)

p	q	$p \wedge q$
T	T	T
T	F	F
F	T	F
F	F	F

(b)

p	q	$p \vee q$
T	T	T
T	F	T
F	T	T
F	F	F

Competency Based Mathematics for Secondary Schools. Book 3

(c)

p	q	q ∧ p
T	T	T
T	F	F
F	T	F
F	F	F

(d)

p	q	q ∧ p
T	T	T
T	F	T
F	T	T
F	F	F

We can see that the truth tables for $p \wedge q$ and $q \wedge p$ are the same.
Therefore,   (i) $p \wedge q \equiv q \wedge p$
Similarly, the truth tables for) $p \vee q$ and $q \vee p$ are the same.
Therefore,   (ii) $p \vee q \equiv q \vee p$

 Investigative Activity

Draw the truth tables for the following.
(a) $\sim p \vee \sim q$ (b) $\sim (p \wedge q)$ (c) $\sim p \wedge \sim q$ (d) $\sim (p \vee q)$.
Deduce that:
(i) $\sim p \vee \sim q \equiv \sim (p \wedge q)$ (ii) $\sim p \wedge \sim q \equiv \sim (p \vee q)$

Solution

(a)

p	q	~p	~q	~p ∨ ~q
T	T	F	F	T
T	F	F	T	T
F	T	T	F	T
F	F	T	T	F

(b)

p	q	~ (p ∧ q)
T	T	T
T	F	T
F	T	T
F	F	F

(c)

p	q	~p	~q	~p∧~q
T	T	F	F	F
T	F	F	T	F
F	T	T	F	F
F	F	T	T	T

(d)

p	q	p∨q	~(p∨q)
T	T	T	F
T	F	F	F
F	T	F	F
F	F	F	T

From the tables we can see that the truth tables for $\sim p \wedge \sim q$ and $\sim(p\vee q)$ are the same.
Therefore,   (i) $\sim p \wedge \sim q \equiv \sim(p\vee q)$
Similarly, the truth tables for $\sim p \vee \sim q$ and $\sim(p\wedge q)$ are the same.
Therefore,   (ii) $\sim p \vee \sim q \equiv \sim(p\wedge q)$

12.13 De Morgan's Laws

The logically equivalent statements in investigative activity above are called the **De Morgan's laws**. These laws are restated below.

> The De Morgan's laws of logic state that:
> (i) $\sim p \wedge \sim q = \sim(p\vee q)$ (ii) $\sim p \vee \sim q = \sim(p\wedge q)$

12.14 Connectors

We call the logical symbols $\sim, \wedge, \vee, \rightarrow$ and $\leftrightarrow$ connectors. Their respective set algebra equivalents are ', $\cap$, $\cup$, $\rightarrow$ and $=$. We call $\sim$ a unitary connector because it affects only one statement and call $\wedge, \vee, \rightarrow$ and $\leftrightarrow$ binary connectors because they combine two statements.

Competency Based Mathematics for Secondary Schools. Book 3

Exercise 12:3

1. Let p be the statement "She is lazy" and q be the statement "She is beautiful". Write down the following using symbolic language.
   (i) She is not lazy.
   (ii) She is beautiful and lazy.
   (iii) She is neither lazy nor beautiful.
   (iv) She is either lazy or beautiful.
   (v) It is not true that she is beautiful and lazy.
   (vi) It is false that she is lazy and beautiful.

2. Decompose the following composite statements into its components.
   (a) Mrs. Ngwa and Mrs. Tayong visited me.
   (b) Nfor likes rice and beans.
   (c) Mr. Nkwain is a Cameroonian ambassador.
   (d) Bamenda and Bafoussam are big cities.

3. Let p and q be two statements defined as follows.
   p: Nfor is hungry.
   q: Nfor is thirsty.
   Write out the following in English.
   (i) $\sim p$ (ii) $p \wedge q$ (iii) $p \vee q$ (iv) $p \Rightarrow q$
   (v) $p \Leftrightarrow q$ (vi) $p \Rightarrow \sim q$ (vii) $\sim p \wedge \sim q$ (viii) $p \Leftrightarrow \sim q$

4. Make a conditional statement by combining the following pair of statements
   (a) Fombe is rich. Fombe is happy.
   (b) He was drunk. He drank alcohol.
   (c) It is night in Cameroon. Places are dark.
   (d) She performed well. She had a prize.

5. Make a biconditional statement by combining the pair of statements in question 4.

6. Draw the truth tables for the propositions
   $(p \rightarrow q) \wedge (q \rightarrow p)$ and $p \leftrightarrow q$.
   Use your tables to determine whether the statement
   $(p \rightarrow q) \wedge (q \rightarrow p) \equiv p \leftrightarrow q$ is true.

7. Draw the truth tables for
   $p \rightarrow q$ and $\sim p \vee q$ and say whether or not $p \rightarrow q \equiv \sim p \vee q$.

8. Proof the following:
   (a) the associative law of logic i.e. $(p \wedge q) \wedge r \equiv p \wedge (q \wedge r)$
   (b) the distributive law of logic i.e. $p \vee (q \wedge r) \equiv (p \vee q) \wedge (p \vee r)$

9. Draw up the truth tables for the following compound statements.
   (a) $(p \vee q) \rightarrow (p \wedge q)$ (b) $(p \vee q) \wedge \sim p$ (c) $p \rightarrow (p \wedge \sim q)$
   (d) $(p \rightarrow q) \wedge q$ (e) $(p \rightarrow q) \rightarrow \sim p$ (f) $p \wedge (p \rightarrow q)$

10. Determine which of the following statements are equivalent.
    (a) $(p \rightarrow q) \wedge r$ and $p \rightarrow (q \wedge r)$
    (b) $p \vee (q \wedge r)$ and $(p \vee q) \wedge (p \vee r)$
    (c) $(p \wedge q) \rightarrow r$ and $p \rightarrow (q \rightarrow r)$

11. Given that water can flow through two taps A and B as shown below. Draw the truth table for A or B using T for open and F for closed.

Module 14, Topic 12: Logic

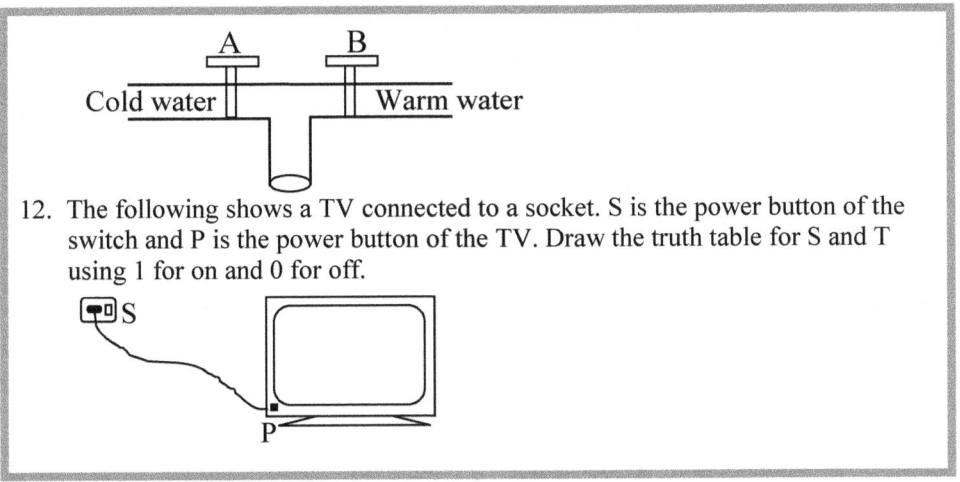

12. The following shows a TV connected to a socket. S is the power button of the switch and P is the power button of the TV. Draw the truth table for S and T using 1 for on and 0 for off.

12.15 Tautologies

A tautology is a proposition, which is always true. Therefore, the last column of the truth table of a tautology is made up of all Ts or 1s. For instance, the proposition "p or not p" is a tautology.

p	$\sim p$	$p \lor \sim p$
T	F	T
F	T	T

12.16 Contradictions

A contradiction is a proposition, which is always false. Therefore, the last column of the truth table of a contradiction is made up of all Fs or 0s. For instance, the proposition "p and not p" is a contradiction.

p	$\sim p$	$p \land \sim p$
T	F	F
F	T	F

12.17 Quantifiers

Logical operators which are used to stand for words such as for all, for every, for some, for each, for any, there exist, for at least etc. are called **Quantifiers** because they suggest the idea of quantity. The common quantifiers are universal, existential and unitary quantifiers.

(a) Universal Quantifier

The symbol for the universal quantifier is ∀, read "for all" or "for every". Thus if $p(x)$ is a true statement concerning all the elements in a set A, we can express this idea symbolically as follows.

$$\forall x \in A, p(x) \text{ or } (\forall x \in A)\, p(x).$$

This is read as "for all x belonging to A, $p(x)$ is true" or "for every element x of A, $p(x)$ is true".

The fundamental property of the universal quantifier is as follows.
(1) True: if for all x in A, $p(x)$ is true.
(2) False: If there is at least one element in A, for which $p(x)$ is false.

(b) Existential Quantifier

The symbol for the existential quantifier is ∃, read "there exists" or "for at least one" or "for some". Thus if $p(x)$ is a true statement concerning at least one but not all the elements in a set A, we can express this idea symbolically as follows.

$$\exists x \in A, p(x) \text{ or } (\exists x \in A)\, p(x).$$

(c) Unitary Existential Quantifier

The symbol for the unitary existential quantifier is ∃! read, "there exists one and only one". Thus if $p(x)$ is a true statement concerning only one element in a set A, we can express this idea symbolically as follows.

$$\exists! x \in A, p(x) \text{ or } (\exists! x \in A)\, p(x).$$

This is read as "there exist one and only one element x belonging to A, for which $p(x)$ is true".

The fundamental property of the unitary existential quantifier is as follows.
(1) True: If there is at least one element x in A, for which $p(x)$ is true.
(2) False: If for all, x in A, $p(x)$ is false.

Example

In each of the following cases write statements using a quantifier and explain each in English.
(1) Let $A = \{1,2,3,4,5\}$ and $p(x)$: $0 \leq x - 1 < 5$.
(2) Let $E = \{2,4,6,8,10\}$ and $p(x)$: x is even.
(3) Let $E = \{2,4,6,8,10\}$ and $p(x)$: x is a multiple of 2.
(4) $\mathbb{R}$ is the set of real numbers and $p(a)$: $a + 1 = 1 + a = a$.

(5) Let $A = \{1,2,3,4,5\}$ and $p(x)$: $x - 1$ is odd.
(6) Let $E = \{2,4,6,8,10\}$ and $p(x)$: x is a multiple of 4.
(7) $\mathbb{R}$ is the set of real numbers and $p(a)$: $a \times 0 = 0 \times a = 0$.
(8) $\mathbb{R}$ is the set of real numbers and $p(a)$: $a \times 1 = 1 \times a = a$.

Solution
(1) $\forall x \in A, p(x)$ or $(\forall x \in A) \, p(x)$.
   This statement means "for all values of x belonging to the set A,
   $0 \leq x - 1 < 5$.
(2) $\forall x \in E, p(x)$ or $(\forall x \in E) \, p(x)$.
   This statement means, "for all values of x belonging to the set E, x is even".
(3) $\forall x \in E, p(x)$ or $(\forall x \in E) \, p(x)$.
   This statement means, "for all values of x belonging to the set E, x is a multiple of 2".
(4) $\forall a \in \mathbb{R}, \sim p(a)$ or $(\forall a \in \mathbb{R}) \sim p(a)$.
   This statement means "for all real values of a, $a + 1 = 1 + a = a$.
(5) $\exists x \in A, p(x)$ or $(\exists x \in A) \, p(x)$.
   This statement means "There exist at least one element x belonging to the set A, for which $p(x)$ is true".
(6) $\exists x \in E, p(x)$ or $(\exists x \in E) \, p(x)$.
   This statement means, "There are some elements in the set E, which are multiples of 4".
(7) $\exists! x \in \mathbb{R}, p(x)$ or $(\exists! x \in \mathbb{R}) \, p(x)$.
   There exists one and only one element a belonging to $\mathbb{R}$, which satisfies the condition $a \times 0 = 0 \times a = 0$.
(8) $\exists! x \in A, p(x)$ or $(\exists! x \in A) \, p(x)$.
   There exists one and only one element belonging to $\mathbb{R}$, which satisfies the condition $a \times 1 = 1 \times a = a$.

Exercise 12:4

Using variables of your choice to represent the sets involved, write out the following using the quantifiers $\forall$, $\exists$ or $\exists!$
1. 2 is the only even prime number.
2. All doctors are scientists.
3. Some footballers play volleyball.
4. No student passed the examination.
5. In $\mathbb{R}$, 0 is the additive inverse for addition.
6. All fishes swim.
7. Some parallelograms are rectangles
8. There is no real number such that $x^2 + 1 = 0$

12.18 Syllogisms

A **syllogism** is an argument made up of statements in one of four forms:

(a) Universal affirmative
    Form: All A's are B's.
    e.g. p: all cows eat grass.
(b) Universal negative
    Form: No A's are B's.
    E.g., No historians are chemists.
(c) A particular affirmative
    Form: Some A's are B's.
    e.g., some polygons are rectangles.
(d) A particular negative
    Form: Some A's are not B's.
    E.g., some students are not studious.

The common nouns, such as cows, grass, historians, chemists etc are called the terms of the syllogism.

12.19 Hypotheses and Conclusions

Consider the following statements.
p: All even numbers are divisible by 2.
q: 14 is an even number
    Therefore,
r: 14 is divisible by 2.

In logic, we call statements such as p and q the **premises** or **hypotheses** and call a statement such as r the **conclusion**.
A well-formed syllogism consists of two premises and a conclusion, each premise having one term in common with the conclusion and one in common with the other premise.
It follows that if the premises or hypotheses are true then the conclusion is bound to be true.

Module 14, Topic 12: Logic

 Exercise 12:5

Write a conclusion r which follows from the given premises p and q.
1. p: All politicians are liars.
   q: Ngoh is a politician
2. p: Some polygons are parallelograms.
   q: Some parallelograms are rectangles.
3. p: $x > 7$
   q: $x < 4$
4. p: The rich are never happy.
   q: Nfor is rich.
5. p: Bamenda is in Cameroon.
   q: Cameroon is in Europe.
6. p: x is a multiple of $4 \Rightarrow x$ is a multiple of 2.
   q: 18 is a multiple of 4.
7. p: If each angle in a quadrilateral is a right angle, then the quadrilateral is a rectangle.
   q: The quadrilateral $ABCD$ is a rectangle.
8. p: All rhombuses have equal sides.
   q: $ABCD$ is not a rhombus.
9. p: The beautiful ones are not yet born.
   q: Bih was born in 1998.
10. p: $A > B$
    q: $B < A$

 Multiple Choice Exercise 12

1. Given the truth table

p	q
T	F
F	T

   The correct relationship between the statements p and q is:
   [A] $p \Rightarrow q$ [B] $p \Leftrightarrow q$ [C] $\sim p = q$ [D] $p = q$
2. The closed statement among the following is:
   [A] $4 + 4 = 8$ [B] He ate rice   [C] $2x - 1 \leq 0$ [D] $2x - 1 = 0$
3. The open statement among the following is:
   [A] $7 + 7 = 14$ [B] Her makeup is good
   [C] $3 + 4 \leq 0$ [D] Ngwa is a footballer
4. Which of the following is not a statement?
   [A] $3p - 7 = 5$. [B] Roses are red.
   [C] Tse is sick. [D] Read your Bible everyday.
5. A statement among the following is:

[A]   5 + 8   [B] Hit the iron while hot

[C] $ax^2 + bx + c$ [D] $x = -b \pm \dfrac{\sqrt{b^2 - 4ac}}{2}$

6. The truth set for the statement $X = \{x: 2x - 1 \leq 7, x \in \mathbb{N}\}$ is:
   [A] $X = \{0, 1, 2, 3, 4\}$ [B] $X = \{x: x < 4, x \in \mathbb{N}\}$
   [C] $X = \{x: 0 \leq x < 4, x \in \mathbb{N}\}$ [D] $X = \{x: 0 \leq x \leq 4, x \in \mathbb{N}\}$

7. A statement among the following is:
   [A] To God be the glory.   [B] 10–3
   [C] Stand up when the visitor arrives   [D] $5x > -20$

8. If p: Bih is in Bamenda.
      q: Bamenda is in Cameroon.
   Then,
   [A] $q \Rightarrow p$ [B] $\sim p \Rightarrow q$ [C] $p \Rightarrow q$ [D] $\sim q \Rightarrow p$

9. Given that,
      p: $\triangle ABC$ is equilateral.
      q: $\triangle ABC$ is equiangular.
   Then,
   [A] $\sim p \Rightarrow q$ [B] $\sim q \Leftrightarrow p$ [C] $p \Leftrightarrow q$ [D] $\sim p \Rightarrow q$

10. The truth table for $p \wedge q$ is:

p	q	$p \wedge q$
T	T	T
T	F	F
F	T	F
F	F	F

[A]

p	q	$p \wedge q$
T	T	T
T	F	T
F	T	T
F	F	F

[B]

p	q	$p \wedge q$
T	T	T
T	F	F
F	T	T
F	F	T

[C]

p	q	$p \wedge q$
T	T	T
T	F	T
F	T	F
F	F	F

[D]

11. The truth table for $p \vee q$ is:

p	q	$p \vee q$
T	T	T
T	F	F
F	T	F
F	F	F

[A]

p	q	$p \vee q$
T	T	T
T	F	T
F	T	T
F	F	F

[B]

p	q	$p \vee q$
T	T	T
T	F	F
F	T	T
F	F	T

[C]

p	q	$p \vee q$
T	T	T
T	F	T
F	T	F
F	F	F

[D]

12. The following statements refer to the figure above.
    $p: a+b = \theta$
    $q: c+\theta = 180°$
    $r: a+b+c = 180°$

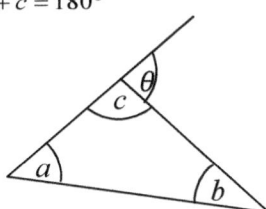

The arguments, which proves that the exterior angle of a triangle is equal to the sum of the two interior opposite angles, is:
[A] $q \Rightarrow r \Rightarrow p$ [B] $r \Rightarrow p \Rightarrow q$ [C] $r \Rightarrow q \Rightarrow p$ [D] $q \Rightarrow p \Rightarrow r$

13. Which of the following is a composite statement?
    [A] He studied Mathematics at E.N.S.   [B] The kola nut has been eaten.
    [C] He can drive.   [D] There is no chalk in the school.

14. Which of the following statements is not a composite statement?
    [A] Bamenda is in the NW region of Cameroon.
    [B] Ngala was dead and buried.
    [C] Fonjong eats cocoyam.
    [D] The sun is shining when it is raining.

15. The truth table that represents a tautology is:

p	q	$p \wedge q$
T	T	T
T	F	T
F	T	T
F	F	T

[A]

p	q	$p \wedge q$
T	T	T
T	F	F
F	T	F
F	F	F

[B]

p	q	$p \wedge q$
T	T	T
T	F	F
F	T	F
F	F	T

[C]

p	q	$p \wedge q$
T	T	F
T	F	F
F	T	F
F	F	F

[D]

16. The truth table that represents a contradiction is:

p	q	$p \wedge q$
T	T	T
T	F	T
F	T	T
F	F	T

[A]

p	q	$p \wedge q$
T	T	T
T	F	F
F	T	F
F	F	F

[B]

p	q	$p \wedge q$
T	T	T
T	F	F
F	T	F
F	F	T

[C]

p	q	$p \wedge q$
T	T	F
T	F	F
F	T	F
F	F	F

[D]

17. An inclusive disjunction among the following is:
    [A] Nursing requires a mastery of physics or chemistry.
    [B] At 6 a.m. I shall go to Bamenda or Douala.
    [C] I shall be sleeping or eating.
    [D] In high school, I shall offer LS1 or LA1.
18. An exclusive disjunction among the following is:
    [A] $-5 < 0$ or $5 > 2$.
    [B] $4 > 1$ or $4 < 6$.
    [C] $\triangle ABC$ is equilateral or a right-angled triangle.
    [D] $0 < x \leq 20$ or $x > 3$.
19. The symbol, which is not a logical connector, is:
    [A] $\wedge$ [B] $>$ [C] $\vee$ [D] $\rightarrow$
20. The unitary connector among the following logical symbols is:
    [A] $\sim$ [B] $>$ [C] $\vee$ [D] $\rightarrow$
21. The statement, which is not a syllogism, is:
    [A] All cows eat grass.   [B] No historians are chemists.
    [C] I love yams.   [D] Some polygons are rectangles.
22. The negation of the statement "All Anglophones speak English" is:
    [A] Some Anglophones speak English.
    [B] Not all Anglophones speak English.
    [C] No Anglophone speaks English.
    [D] All Anglophones do not speak English.
23. The negation of the statement "All x are not y" is:
    [A] Some x are y.   [B] Some x are not y.
    [C] No x are y.   [D] All x are not y.
24. The negation of the statement "Not all that glitters is gold" is:
    [A] All that glitters is not gold.   [B] Some of what glitters is gold.
    [C] Nothing that glitters is gold.   [D] All that glitters is gold.
25. Given the following statements:
    p: It is raining.
    q: I will go to school by car.
    The statement $p \wedge q$ is:
    [A] It is raining or I will go to school by car.
    [B] It is raining and I will go to school by car.
    [C] It is raining and if I will go to school by car.
    [C] It is raining and if and only if I will go to school by car.

Topic 13

TRANSPOSITION OF FORMULAE

Objectives
At the end of this topic, the learner should be able to:

1. Make a subject of a formula, which does not contain square roots.
2. Make a subject of a formula, which contains square roots.
3. Make a subject of a formula, which involves quadratics.

13.1 Making a Subject of a Formula

The methods used in making a subject of a formula are the same as those used in solving equations. The only difference is that the subject will be in terms of the other unknowns in the formula.

13.2 Formulae without Square Roots

Solve for the subject in the same way as equations are solved. In some cases, factorization is required.

Example

1. Make x the subject of the formula $Cx - R = T$.

    Solution
    $$Cx - R = T$$
    Adding R to both sides
    $$Cx = T + R$$
    Dividing both sides by C
    $$x = \frac{T+R}{C}$$

2. Make y the subject of the formula $a(y + B) = C$.

    Solution
    $$a(y + B) = C$$
    Expanding the left hand side.
    $$ay + aB = C$$
    Subtracting aB from both sides
    $$ay = C - aB$$
    Dividing both sides by a
    $$y = \frac{C - aB}{a} \text{ or } y = \frac{C}{a} - B$$

3. Make n the subject of the formula $f = ea - nh$.

    Solution
    $$f + nh = ea$$

Module 14, Topic 13: Transposition of Formulae

$$nh = ea - f \Leftrightarrow n = \frac{ea - f}{h}$$

4. Make a the subject of the formula $\dfrac{t}{a} = \dfrac{b}{e}$.

   Solution
   $$\frac{t}{a} = \frac{b}{e}$$
   $$et = ab \Leftrightarrow a = \frac{et}{b}$$

5. Make x the subject of the formula $r - \dfrac{m}{x} = p^2$.

   Solution
   $$r - \frac{m}{x} = p^2$$
   $$rx - m = p^2 x$$
   $$rx - p^2 x - m = 0$$
   $$rx - p^2 x = m$$
   $$x(r - p^2) = m \Leftrightarrow x = \frac{m}{r - p^2}$$

6. Make y the subject of each of the following formulae.

   (a) $Ny + B = C - Ny$ (b) $\dfrac{p - y}{p + y} = q$

   Solution
   (a) $Ny + B = C - Ny$
   $$2Ny + B = C$$
   $$2Ny = C - B \Leftrightarrow y = \frac{C - B}{2N}$$

   (b) $\dfrac{p - y}{p + y} = q$
   $$p - y = q(p + y)$$
   $$p - y - pq = qy$$
   $$p - pq = qy + y$$
   $$p - pq = y(q + 1) \Leftrightarrow y = \frac{p - pq}{q + 1}$$

Competency Based Mathematics for Secondary Schools. Book 3

Exercise 13:1

1. Express w in terms of a, b, u and T given that $T = \dfrac{wa}{(u+w)b}$.

2. The formula for converting temperatures from Celsius ($C°$) to Fahrenheit ($F°$) is given by $F = \dfrac{9}{5}C + 32$. Find the temperature at which the Celsius and Fahrenheit thermometers record the same values. Make C the subject.

3. Given that $S = \dfrac{T}{5} - \dfrac{2R}{W}$. Make W the subject of the formula and find W when $T = 10.5$, $S = 1$ and $R = 5.5$.

4. Given that $V = \dfrac{\pi r^2 h}{c}$.
    (a) Find the numerical value of V, giving your answer in standard form, given that $r = 0.01$, $c = 528$, $h = 25.2$ and $\pi = \dfrac{22}{7}$.
    (b) Express h in terms of V, c, r, and π.

5. Given that $\dfrac{2x+y}{5y+3x} = \dfrac{1}{4}$. Find the value $\dfrac{x}{y}$.

6. Express p in terms of m and n given that $\dfrac{n}{4} = \dfrac{m}{2} - \dfrac{p}{8}$.

7. (i) In the formula $L = \dfrac{gh}{(g-h)k}$, make g the subject.
   (ii) Find the value of k if $g = 12$, $h = 4$ and $L = 3$.

13.3 Formulae Containing Square Roots

If there is a square root sign, first square both sides to eliminate the square root sign.

Example

Make x the subject of each of the following formulae.
(a) $\sqrt{Ax + B} = \sqrt{C}$
(b) $t = \sqrt{m + x^2}$
(c) $a\sqrt{k^2 - x} = b$
(d) $\sqrt{\left(\dfrac{a}{x} - b\right)} = c$

240

Module 14, Topic 13: Transposition of Formulae

Solution

(a) $\sqrt{Ax+B} = \sqrt{C}$
Squaring both sides
$Ax + B = C$
$Ax = C - B \Leftrightarrow x = \dfrac{C-B}{A}$

(b) $t = \sqrt{m + x^2}$
Squaring both sides
$t^2 = m + x^2$
$t^2 - m = x^2 \Leftrightarrow x = \pm\sqrt{t^2 - m}$
Remember the $\pm$ sign or else a solution is lost.

(c) $a\sqrt{k^2 - x} = b$
Squaring both sides and expanding.
$a^2(k^2 - x) = b^2$
$a^2 k^2 - a^2 x = b^2$
$-a^2 x = b^2 - a^2 k^2 \Leftrightarrow x = \dfrac{a^2 k^2 - b^2}{a^2}$

(d) $\sqrt{\left(\dfrac{a}{x} - b\right)} = c$
Squaring both sides
$\dfrac{a}{x} - b = c^2$
$a - bx = c^2 x$
$a = (c^2 + b)x$
$\Leftrightarrow x = \dfrac{a}{(c^2 + b)}$

Exercise 13:2

1. Given that $V = \dfrac{\pi n}{\sqrt{(A-3)}}$, express A in terms of V, π and n.

2. Given that $a = p\left(1 + \sqrt{\dfrac{r}{t}}\right)$, express t in terms of a, p and r.

3. Make p the subject of the formula $m = 2n\pi \sqrt{\left(\dfrac{k+1}{p}\right)}$.

   Hence, or otherwise, and taking π as $\dfrac{22}{7}$, find the value of p when $m = 22$, $n = 7$ and $k = 15$.

4. Given that $p = \sqrt{\dfrac{nq}{na+b}}$, make n the subject of the formula.

5. (a) Make v the subject of the formula $y = \dfrac{k}{\sqrt[n]{v}}$.

   (b) Given that $k = 16$ and $y = 4$, find v when $n = \dfrac{1}{2}$.

6. Make l the subject of the formula $T = \dfrac{1}{2\pi}\sqrt{\dfrac{l}{g}}$.

7. Make a the subject of the formula $\sqrt{\dfrac{ra = t}{s}} = v$.

8. Given that, $p = \sqrt{\dfrac{m+n}{m}}$ express m in terms of p and n.

9. The formula for finding the volume of a certain composite figure is $V = \dfrac{1}{3}\pi r^2 (h+r)$. Make h the subject of the formula.

13.4 Formulae Containing Quadratics

When a formula contains a quadratic, the subject often has two different values. To find the two values use the factorisation method, the quadratic formula or the method of completing the square.

 Example

Make x the subject of the formula $2x^2 + 7xyz - 15y^2z^2 = 0$. Find the values of x for which $y = 2$ and $z = -7$

Solution

$$2x^2 + 7xyz - 15y^2z^2 = 0$$

Module 14, Topic 13: Transposition of Formulae

$$2x^2 + 7x(yz) - 15(yz)^2 = 0$$
$$(2x - 3yz)(x + 5yz) = 0$$
$$x = \frac{3}{2}yz \text{ or } x = -5yz$$

Alternatively, the quadratic formula $x = \dfrac{-b \pm \sqrt{b^2 - 4ac}}{2a}$ may be used.

Thus, $a = 2, b = 7yz, c = -15y^2z^2$

$$x = \frac{-7yz \pm \sqrt{(7yz)^2 - 4(2)(-15y^2z^2)}}{2(2)}$$

$$x = \frac{-7yz \pm \sqrt{169y^2z^2}}{2(2)}$$

$$x = \frac{-7yz \pm 13yz}{4}$$

$$x = \frac{3yz}{2} \text{ or } x = -5yz$$

When $y = 2$ and $z = -7$

$$x = \frac{3}{2}(2)(-7) \text{ or } x = -5(2)(-7)$$

$$\Leftrightarrow x = -21 \text{ or } x = 70$$

Exercise 13:3

1. Make the letter in bracket a subject.
   (a) $A = \pi rl + \pi r^2$ (r)
   (b) $S = ut + \dfrac{1}{2}at^2$ (t)
   (c) $ax^2 + c = bx$ (x)
   (d) $A = \pi\{r + t\}^2 - r^2$ (t)

2. Given that $r^2 + 3s^2 = 4rs$, find the two values of $\dfrac{r}{s}$.

   In the problem 3 to 5, make the letter in bracket the subject of the given formula.

3. $T^2 + TG = 12G^2$ (G)
4. $6p^2 + 17pqr = 3q^2r^2$ (r)
5. $12m^2 - 16mn + 5n^2 = 0$ (n)

Competency Based Mathematics for Secondary Schools. Book 3

 Multiple Choice Exercise 13

1. Given that $v = u + at$. In terms of u, v and t, a will be:

   [A] $\dfrac{v+u}{t}$ [B] $\dfrac{v+t}{u}$ [C] $\dfrac{v-u}{t}$ [D] $\dfrac{t+u}{v}$

2. The formula for finding the sum of the interior angles of a polygon of n sides is $S = 180(n - 2)$. The number of sides in that polygon is:

   [A] $n = \dfrac{S}{180} - 2$ [B] $n = \dfrac{S}{180}$ [C] $n = \dfrac{2s}{180}$ [D] $n = \dfrac{S}{180} + 2$

3. The volume V of a cylinder is $V = \pi r^2 h$ where r and h are the radius and height of the cylinder respectively. The height h is:

   [A] $h = V\pi r^2$ [B] $\dfrac{V}{\pi r^2}$ [C] $\dfrac{\pi r^2}{V}$ [D] $\dfrac{\pi}{Vr^2}$

4. As a subject of the formula $A = 2\pi rh$, r is:

   [A] $\dfrac{A}{2\pi h}$ [B] $\dfrac{\pi}{2Ah}$ [C] $\dfrac{\pi h}{2A}$ [D] $2Ah\pi$

5. The formula for finding the volume of a cone is $V = \tfrac{1}{3}\pi r^2 h$. In terms of v, r and π, the height h is:

   [A] $\dfrac{\pi r^2}{3}$ [B] $\dfrac{3Vr^2}{\pi}$ [C] $\dfrac{3V}{\pi r^2}$ [D] $\dfrac{\pi^2 r}{3}$

6. In the temperature conversion formula $F = \dfrac{9}{5}C + 32$, C is:

   [A] $\dfrac{5F - 160}{9}$ [B] $\dfrac{5F + 160}{9}$ [C] $\dfrac{9F + 160}{5}$ [D] $\dfrac{9F - 160}{9}$

7. If $I = \dfrac{PRT}{100}$, the value of T when $P = 450$, $R = 12$ and $I = 90$ is:

   [A] $\dfrac{3}{5}$ [B] $\dfrac{5}{6}$ [C] $\dfrac{5}{3}$ [D] 15

8. If $y = \dfrac{a+p}{a-p}$, then :

   [A] $\dfrac{2a-y}{a+y} = p$ [B] $\dfrac{ay-1}{y+1} = p$ [C] $\dfrac{a(y-1)}{y+1} = p$ [D] $\dfrac{2y-1}{y-1} = p$

9. If $h(m+n) = m(h+r)$, h in terms of m, n and r is:

Module 14, Topic 13: Transposition of Formulae

[A] $h = \dfrac{mr}{2m+n}$ [B] $h = \dfrac{mr}{2m-n}$ [C] $h = \dfrac{m+r}{n}$ [D] $h = \dfrac{mr}{n}$

10. If we make q the subject of the relation $t = \sqrt{\dfrac{pq}{r} - r^3 q}$, the relation will now be:

[A] $q = \dfrac{t^2}{p - r^4}$ [B] $q = \dfrac{rt^2}{p - r^4}$ [C] $q = \dfrac{rt}{p - r^4}$ [D] $q = \dfrac{p - r^4}{rt^2}$

11. Making S the subject of the formula $V = \dfrac{K}{\sqrt{T - S}}$ gives:

[A] $S = T - \dfrac{K^2}{V^2}$ [B] $S = \dfrac{K^2}{V^2} - T$ [C] $S = T - \dfrac{V^2}{K^2}$ [D] $S = T\left(\dfrac{V^2 - K^2}{V^2}\right)$

12. If $y = \sqrt{ax - b}$, x in terms of y, a and b is:

[A] $x = \dfrac{y^2 - b}{a}$ [B] $x = \dfrac{y + b}{a}$ [C] $x = \dfrac{y - b}{a}$ [D] $x = \dfrac{y^2 + b}{a}$

13. $k = m\sqrt{\dfrac{t - p}{r}}$, as subject of the formula, t is equal to:

[A] $\dfrac{rk^2 + p}{m^2}$ [B] $\dfrac{rk^2 + pm^2}{m^2}$ [C] $\dfrac{rk^2 - p}{m^2}$ [D] $\dfrac{rk^2 - p^2}{m^2}$

14. Given that $U = \dfrac{T}{5} - \dfrac{2R}{Q}$, then T expressed in terms of U, R and Q is:

[A] $T = 5U + \dfrac{2R}{Q}$ [B] $T = 5U + \dfrac{10R}{Q}$ [C] $T = 5\left(U - \dfrac{2R}{Q}\right)$ [D] $T = 5U - \dfrac{2R}{Q}$

Topic 14

VARIATION

Objectives
At the end of this topic, the learner should be able to:

1. State whether two variables are in direct or inverse variation.
2. Interpret, write and use the symbol $\propto$.
3. Solve problems on direct or inverse variation.
4. Interpreted and Solve simple problems on partial and joined Variation.

Module 14, Topic 14: Variation

14.1 Real Life Examples of Variation

In real life, there are many related quantities, in one way or the other. Examples of such related quantities are:
1. The interest generated per year increases as the amount of money invested increases.
2. The time taken to complete a given piece of work reduces, as the number of people doing the work increase.
3. The electric bill increases as the amount of current used increases.
4. The area of the circle of a circle increases as the radius increases.
5. Temperature decreases with increasing altitude (i.e. the higher you go, the colder it becomes)
6. The amount of current passing in a wire increases with decreases in resistance.

The above are examples of variations and in each case; it is possible to find a mathematical relation or formula connecting the two quantities. There are two basic types of variation.

- Direct variation or direct proportion and
- Inverse variation or inverse proportion.

14.2 Direct variation

Suppose that one pen cost 100 FRS. The following table shows the prices of 1,2,3 ... 10 pens.

Number of pens, n	Cost of pens, FRS
1	100
2	200
3	300
4	400
5	500
6	600
7	700
8	800
9	900

Notice that the number of pens increases with the price. This is an example of direct variation or direct proportion.

If two quantities x and y are related in this way, then y is said to be **directly proportional** to x or y varies directly as x. Direct variation be expressed mathematically as

$$y \propto x \Rightarrow y = kx \Leftrightarrow \frac{y}{x} = k$$

We call k is a **constant of proportionality**. We can find this constant if we know the value y for a given value of x.

Thus, when $y = 12$ and $x = 2$, $k = \frac{12}{2} = 6$

 Example

1. Given that y is directly proportional to x and $y = 64$ when $x = 8$. Find the value of y when $x = 20$.

   Solution

   $$y \propto x \Rightarrow k = \frac{y}{x} \Rightarrow k = \frac{64}{8} = 8$$

   When $x = 20$, $y = kx \Rightarrow y = 8(20) = 160$

2. Two quantities x and y are such that y varies directly as x. Copy and complete the following table, hence draw a graph to represent this variation. Use your graph to find the value of y when $x = 2.5$.

x	0	5	10	15	20	25
y				3		

   Solution
   From the table, $x = 15$ when $y = 3$
   $$\Rightarrow k = \frac{y}{x} = \frac{3}{15} = \frac{1}{5}.$$
   So $y = kx = \frac{1}{5}x$ and the complete Table is

x	0	5	10	15	20	25
y	0	1	2	3	4	5

   The following figure shows the graph. Notice that, the graph is a straight line passing through the origin. This is the usual nature of a graph of direct variation.

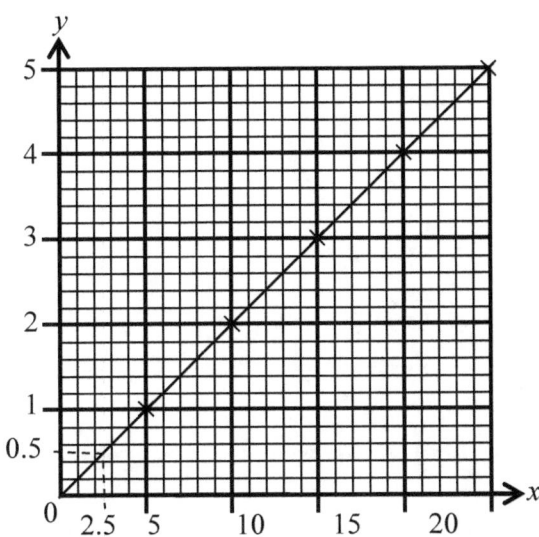

There are so many other cases of direct variation. For instance, one quantity may vary directly as the square or square root of the other. Thus, the area of a circle varies directly as the square of its radius.

3. The variables x and y are known to be related by the formula $y = kx^2$. Given that $x = -2$ when $y = 1$, find the value of k and hence calculate the value x when $y = 16$. Also, make a table of values of x against y for integral values of x in the range $-4 \leq x \leq 4$.

Use this table to draw the graph of y against x.

Solution

$$k = \frac{y}{x^2} = \frac{1}{(-2)^2} = \frac{1}{4} \Rightarrow y = \frac{1}{4}x^2$$

But $kx^2 = y \Rightarrow x = \pm\sqrt{\frac{y}{k}}$

When $y = 16$, $x = \pm\sqrt{16 \div \frac{1}{4}} = \pm 8$

x	−4	−3	−2	−1	0	1	2	3	4
y	4	2.25	1	0.25	0	0.25	1	2.25	4

The graph below is the typical shape of the graph of $y = kx^2$. Notice that the function is a quadratic function with the coefficient of x and the independent term both zero.

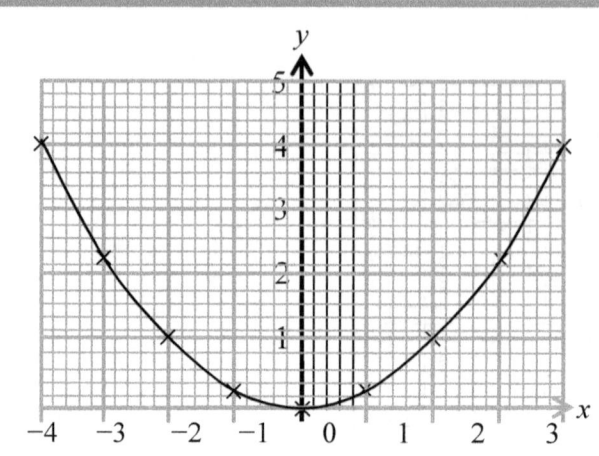

 Exercise 14:1

1. Given that $c \propto n$ and $c = 28$ when $n = 4$. Find the formula connecting c and n.
2. Given that m is directly proportional to n and k is the constant of proportionality. Find a formula connecting m and n.
3. Given that x varies directly with y and $x = 7$ when $y = 35$. Find the relationship between x and y.
4. Given that $x \propto y$ and when $x = 4$, $y = 20$. Find x when $y = 5$.
5. Given that t varies directly as the square of m and $t = 10$ when $m = 25$. Calculate t when $m = 4$.
6. Given that x varies directly as the square of y and $x = 4$ when $y = 6$. Find the value of y when $x = 16$.
7. Given that s varies directly as T^2 and $T = 2$ cm when $s = 12$ cm^2. Find s when $T = 3$ cm.
8. Given that p varies directly as q and $p = 4.5$ when $q = 12$. Find the relationship between p and q and hence find p when $q = 16$.
9. Given that $y \propto x^2$ and that $y = 54$ when $x = 3$, find the constant of proportionality hence the value of x when $y = 24$.
10. Given the mapping

    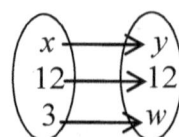

    Find the value of w given that y varies directly as the square of x.

14.3 Inverse variation

Suppose that one person can take 60 minutes to fill a tank with water. The following table shows the amount of time (t) in minutes that 1,2,3,4,5 and 6 people all working at the same rate can use to fill a similar tank.

Number of People, P	1	2	3	4	5	6
Time taken, t (mins)	60	30	20	15	12	10

Notice that the time decreases with the number of people. This is an example of inverse variation or inverse proportion described by 'y is inversely proportional to x' or 'y varies inversely with x' and symbolically expressed as

$$y \propto \frac{1}{x} \Leftrightarrow y = \frac{k}{x} \Leftrightarrow k = yx,$$

Where, k is the constant of proportionality.
Thus, when $x = 2$ and $y = 5$, $k = 2(5) = 10$.

Example

1. Given that y is inversely proportional to x and $y = 4$ when $x = 12$ find the value of y when $x = 3$.

   Solution

   $$y \propto \frac{1}{x} \Leftrightarrow k = yx$$

   When $y = 4$ and $x = 12, k = 4(12) = 48$.
   When $x = 4, y = \frac{k}{x} = \frac{48}{3} \Rightarrow y = 16$.

2. Two quantities x and y are such that y varies inversely as x. Copy and complete the following table then, draw a graph to represent the variation. Use your graph to find the value of y when $x = 1.5$.

x	−4	−3	−2	−1	0	1	2	3	4
y				−12					3

   Solution
   From the table, $x = 4$ when $y = 3$.
   $$y = \frac{k}{x} \Rightarrow k = yx = 4(3) = 12 \Rightarrow y = \frac{12}{x}.$$

The following is the required table.

x	−4	−3	−2	−1	0	1	2	3	4
y	−3	−4	−6	−12	∞	12	6	4	3

The graph is as follows.

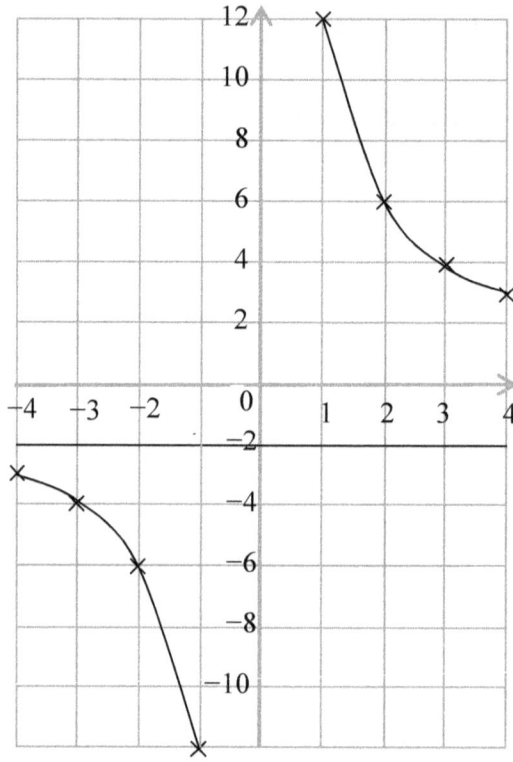

From the graph, when $x = 1.5, y = 8$.

We call this type of graph a **rectangular hyperbola**. The graph shows the typical nature of a graph of inverse variation.

Clearly, the curve approaches the x and the y-axes but never touching it. This is because neither x nor y can be exactly zero.
When $y = 0, x = \dfrac{k}{0} = \infty$ and when $x = 0$,
$y = \dfrac{k}{0} = \infty$.
When a curve approaches a line in this way but never touching it, we call such a line an **asymptote** of the curve. Therefore, the x and the y-axes are the asymptotes of the above curve.

Module 14, Topic 14: *Variation*

14.4 Other Inverse Variations

Consider the statement "*y* is inversely proportional to the square of *x*". We can express this statement symbolically as;

$$y \propto \frac{1}{x^2} \Rightarrow y = \frac{k}{x^2} \text{ or } k = yx^2$$

On the other hand, we can express the statement "*y* is inversely proportional to the square root of *x*, as

$$y \propto \frac{1}{\sqrt{x}} \Rightarrow y = \frac{k}{\sqrt{x}} \text{ or } k = y\sqrt{x}$$

Example

The variables *x* and *y* are known to be related by the formula $x^2 y = k$. Given that $x = -2$ when $y = 36$, find the value of *k* and hence make a table of values of *y* against *x* for $-4 \leq x \leq 4$. Use this table to draw a graph of *y* against *x*.

Solution

$$k = x^2 y = (-2)^2 (36) = 144 \Rightarrow y = \frac{144}{x^2}$$

The following is the required table.

x	−4	−3	−2	−1	0	1	2	3	4
y	9	16	36	144	∞	144	36	16	9

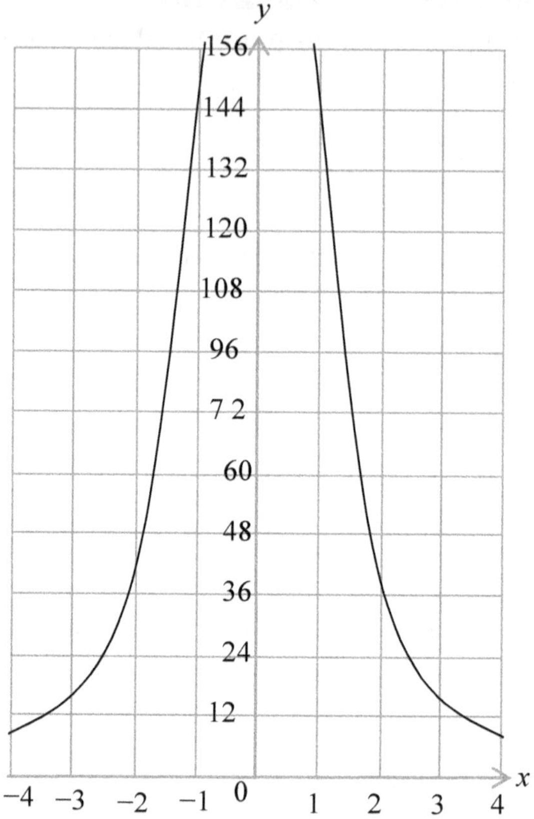

Again, the y-axis and the x-axis are asymptotes to the curve.

Exercise 14:2

1. Given that y is inversely proportional to the square of x and that when $x = \frac{1}{2}, y = 8$.
   (a) Write down an expression to show the relationship between x and y.
   (b) Calculate the value of y when $x = 2\sqrt{3}$.
2. Given that y varies inversely as the square of x, find the numerical values of a and b in the following table.

x	a	2	8
y	16	144	b

Module 14, Topic 14: Variation

3. In the following table, y varies inversely as x. Find the numerical values of a and b.

x	2	4	...	b
y	6	a	...	1

4. Two quantities x and y are related by the $y \propto \dfrac{1}{x^2}$. In an experiment, x and y were measured and the results in the following table were obtained.

x	2	3	4	6	8	12
y	36	16	9	4	3	1

Given that, one of the values of y is wrong. Find the wrong one and give the correct value.

5. Given that x varies as $\dfrac{1}{\sqrt{y}}$ and $y = 4$ when $x = 4$, find

   (a) x when $y = 16$ (b) y when $x = \dfrac{1}{2}$.

6. Given that y varies inversely as the square of x, and that $x = 2$ when $y = 3$. Find the value of x when $y = 27$.

7. In the following table, find the values of a and b if s is inversely proportional to t.

t	5	2	b
s	4	a	144

8. x and y are known to be related by the formula $x^2 y = 16$. In the following table, find which of the value(s) of y is/are incorrect and give the correct value.

x	−2	−1	1	2	3	4
y	4	16	16	4	2	1

9. Given that x varies inversely as $\dfrac{1}{y}$, complete the following table.

x	0	5	10	15	20	25
y				3		

14.5 Joint or Combined Variation

A **joint** or **combined variation** is a variation in which one quantity varies as more than one quantity varies.

Example

1. Write the following statement symbolically. The volume of a right circular cylinder is proportional to both the square of the radius of the cross-section and the height of the cylinder.

   Solution

   $V \propto r^2$ and $V \propto h$

   Combining gives $V \propto r^2 h$

   Introducing the constant of proportionality, k leads to $V = kr^2 h$.

   Mathematicians have proven that this constant is equal to π, hence $V = \pi r^2 h$.

2. The pressure of a fixed mass of gas varies directly as the temperature and inversely as the volume of the gas. Express this statement symbolically.

   Solution

   $P \propto T$ and $P \propto \dfrac{1}{V}$

   Combining gives $P \propto \dfrac{T}{V}$.

   Since the mass, m of the gas is fixed; it means the constant of proportionality is m, $P = \dfrac{mT}{V}$.

3. The height of a cone varies directly as its volume and inversely as the square of its radius. Using k as the constant of proportionality, deduce the formula for calculating the height of a cone.

   Solution

   $h \propto V$ and $h \propto \dfrac{1}{r^2}$

   Combining and introducing the constant of proportionality leads to $h = \dfrac{kV}{r^2}$.

Module 14, Topic 14: Variation

 Exercise 14:3

1. Given that l varies jointly as m and n and that when $m = 2$, $n = 3$ and $l = 9$. Find the law connecting l, m, and n.
2. A is proportional to B and is inversely proportional to C. When $A = 8$ and $B = 4$, $C = 3$.
   (a) Find the formula that connects A, B and C.
   (b) Find A when $B = 5$ and $C = 6$.
3. Given that x varies directly as y and inversely as z and that when $y = 7$ and $z = 3$, $x = 42$. Find the relationship between x, y and z. Find x when $y = 5$ and $z = 9$.
4. Given that x is directly proportional to y and inversely proportional to z and that when $x = 9$, $y = 24$ and $z = 8$. What is the value of x when $y = 5$ and $z = 6$?

 Multiple Choice Exercise 14

1. Given that n varies directly as m and if $n = 8$ when $m = 20$. The value of m when $n = 7$ is:

   [A] 13   [B] 15   [C] $17\frac{1}{2}$ [D] $18\frac{1}{2}$

2. Given that $(x+3)$ varies directly as y and $x = 3$ when $y = 12$, the value of x when $y = 8$ is:

   [A] 1   [B] $\frac{1}{2}$ [C] $-\frac{1}{2}$ [D] -1

3. Given that y is directly proportional to x^2 and $y = 5$ when $x = 2$, then when $x = 6$, $y =$:
   [A] 18   [B] 21   [C] 27   [D] 45
4. P varies inversely as the square of W. When $W = 4$, $P = 9$, then the value of P when $W = 9$ is:

   [A] $\frac{4}{3}$ [B] 6   [C] 4   [D] $\frac{16}{9}$

5. Given that y varies inversely as x^2 then x varies:
   [A] inversely as y^2 [B] inversely as $\sqrt{y}$
   [C] directly as y^2 [D] directly as $\sqrt{y}$

6. Given that x varies inversely as y and $x = \frac{2}{3}$ when $y = 9$, the value of y when

$x = \frac{3}{4}$ is:

[A] $\frac{1}{18}$ [B] $\frac{81}{8}$ [C] $\frac{9}{2}$ [D] 8

7. Given that $y \propto \frac{1}{\sqrt{x}}$ and $x = 16$ when $y = 2$ when $y = 24$, x will be:

   [A] $\frac{1}{9}$ [B] $\frac{1}{6}$ [C] $\frac{1}{3}$ [D] $\frac{2}{3}$

8. Given that x is inversely proportional to m^2 and $x = 3$ when $m = 9$, the value of x when $m = 3$ is:
   [A] 3 [B] 6 [C] 9 [D] 27

9. Given that $p \propto \frac{1}{\sqrt{r}}$ and $p = 3$ when $r = 16$ the value of r when $p = \frac{3}{2}$ is:
   [A] 48 [B] 72 [C] 64 [D] 324

10. Given that R is inversely proportional to S and $R = 15$ when $S = 12$. The value of S when $R = 60$ is:

    [A] $\frac{1}{4}$ [B] 3 [C] 4 [D] 5

11. m varies directly as n and inversely as the square of p; Given that $m = 3$, when $n = 2$ and $P = 1$. The value of m in terms of n and p is:

    [A] $m = \frac{2n}{3p}$ [B] $m = \frac{3n}{2p}$

    [C] $m = \frac{2n}{3p^2}$ [D] $m = \frac{3n}{2p^2}$

12. Given that p varies directly as q while q varies inversely as r. The statement, which is true:
    [A] r varies directly as p. [B] p varies inversely as r.
    [C] p varies directly as r. [D] q varies inversely as p.

13. Given that, 20 men take 6 days to clear a field. The time it would take 12 men working at the same rate to clear a similar field is:
    [A] 40 days [B] 2 days
    [C] $3\frac{1}{2}$ days [D] 10 days

14. K varies directly as N and inversely as the square of L. Given that $L = 1$ when $N = 3$ and $K = 2$. The value of K in terms of N and L is:

[A] $K = \dfrac{2N}{3L^2}$ [B] $K = \dfrac{3N}{2L^2}$

[C] $K = \dfrac{2L^2}{3N}$ [D] $K = \dfrac{3L^2}{2N}$

15. Given that $x \propto y$ and $y \propto \dfrac{1}{z^2}$. The way x varies with z is:

[A] $x \propto \dfrac{1}{z}$ [B] $x \propto \dfrac{1}{\sqrt{z}}$

[C] $x \propto \dfrac{1}{z^2}$ [D] $x \propto z^2$

16. $x \propto y$ and that $x = 28$ when $y = 4$. The formula connecting x and y is:
 [A] $x = 2y$ [B] $x = 4y$
 [C] $x = 7y$ [D] $x = 14y$

17. $x \propto y$ and when $x = 4$, $y = 20$. The value of x when $y = 5$ is:
 [A] 4 [B] 3 [C] 2 [D] 1

18. $m \propto \dfrac{1}{n}$ and $m = 3$ when $n = 2$. The law connecting m and n is:

 [A] $m = 6n$ [B] $m = 3n$ [C] $m = \dfrac{6}{n}$ [D] $\dfrac{3}{n}$

19. Given that $x \propto \dfrac{1}{y}$ and that $x = 9$ when $x = 4$. The formula, which connects, x and y is:

[A] $x = \dfrac{36}{y}$ [B] $x = \dfrac{13}{y}$

[C] $x = \dfrac{9}{y}$ [D] $x = \dfrac{5}{y}$

20. L varies jointly as M and N. When $M = 2$, $N = 3$ and $L = 9$. The law connecting L, M and N is:

[A] $M = \dfrac{2}{3}LN$ [B] $L = MN$

[C] $L = \dfrac{2}{3}MN$ [D] $M = \dfrac{3}{2}LN$

21. R varies directly as t and inversely as m. K is the constant of proportionality. The relationship between R, t and m is:

[A] $R = \dfrac{Km}{t}$ [B] $R = \dfrac{Kt}{m}$

[C] $R = K + \dfrac{m}{t}$ [D] $R = t + \dfrac{K}{m}$

22. The energy E of a moving body varies partly as the square of the height, H of the body above sea level and partly as the square root of its velocity, V. Given that a and b are constants, the equation representing the above expression is:

    [A] $E = aH^2 + b\sqrt{V}$ [B] $E = a\sqrt{H} + bV^2$

    [C] $E = \dfrac{a}{H^2} + \dfrac{b}{\sqrt{V}}$ [D] $E = \dfrac{a}{\sqrt{H}} + \dfrac{b}{V^2}$

23. The equation, which represents the relation in the table below is:

    [A] $y = -3x + 8$ [B] $y = \dfrac{1}{3}x + 8$

    [C] $y = -\dfrac{1}{3}x + 8$ [D] $y = 3x + 8$

x	-3	-2	-1	0	1
y	1	-2	-5	-8	-11

24. The equation, which represents the relation in the table below is:

    [A] $y = 8x + 9$ [B] $y = \dfrac{1}{8}x + 9$

    [C] $y = 8x + 9$ [D] $y = -\dfrac{1}{8}x + 9$

x	-2	-1	0	1	2
y	-7	1	9	17	25

25. The equation, which best describes the relation between x and y in the table below is:

    [A] $y = x^2 + 5$ [B] $y = -x^2 - 5$ [C] $y = x^2 - 5$ [D] $y = -x^2 + 5$

x	-1	0	1	2	3
y	4	5	4	1	-4

Topic 15

RELATIONS AND FUNCTIONS

Objectives

At the end of this topic, the learner should be able to:

1. Appreciate the concept of a relation.
2. Find the Cartesian product of two given sets.
3. State the domain and range of a mathematical relation.
4. Identify the different notations used to denote relations.
5. Define or redefine relations using set builder notation, table of values, sets of ordered pairs, formulas, rule definition, graphs and arrow diagrams.
6. State the inverse of a given relation and represent it on graphs and arrow diagrams.
7. State and use the properties of relations in a set.
8. Determine whether or not a relation is an equivalence relation.
9. Determine whether or not a given relation is a function.
10. Identify functions written using various functional notations.
11. Representation functions on graphs and arrow diagrams.
12. Use flow charts to analyze functions.
13. Identify the different types of mappings and represent them graphically or using arrow diagrams.
14. Analyze a given function using flow charts.
15. Find the inverse of a given rational function.
16. Find the composite function of two or more functions.

RELATIONS

15.1 The Idea of a Relation

Suppose that Mr. Ngala from family A is married to Mafor who is a sister to Mr. Ndoh of family B. Then we can make the following statements concerning these three people.
(i) Mr. Ngala is the husband of Mafor.
(ii) Mr. Ngala is a brother-in-law to Mr. Ndoh.
(iii) Mr. Ndoh is the brother of Mafor.

Statement (i) and (ii) relate Mr. Ngala who is from family A to Mafor and Mr. Ndoh who are from family B. statement (iii) relates Mr. Ndoh to Mafor who are both from family B. Statements (i) and (ii) illustrate relations from one family or set A called the **domain** to another family or set B called the **codomain**. On the other hand, statement (iii) illustrates a relation within only one set called a **relation in a set**.

15.2 The Cartesian product of Two Sets

The Cartesian product of two sets A and B is denoted by $A \times B$, read "A cross B" and is the set of all ordered pairs (a, b), formed by choosing the first element from the set A and the second element from the set B.

$$A \times B = \{(a,b) : a \in A \text{ and } b \in B\}$$

Example

Given that $X = \{x, y, z\}$ and $Y = \{a, b\}$. Find the Cartesian products
(a) $X \times Y$ (b) $Y \times X$

Solution
(a) $X \times Y = \{(x, a), (x, b), (y, a), (y, b), (z, a), (z, b)\}$
(b) $Y \times X = \{(a, x), (a, y), (a, z), (b, x), (b, y), (b, z)\}$

Again, the ordered pair (a, b) should not be confused with the same notation used to represent an open interval.

Investigative Activity

1. Given that $A = \{1,2,3\}$ and $B = \{x, y\}$. Compute the Cartesian products
   (a) $A \times B$ (b) $B \times A$
2. Is the Cartesian product commutative?

From the above investigation, we see that $A \times B \neq B \times A$. Hence, the Cartesian product is not commutative. This means that in an ordered pair (a, b), the order of the elements is very important. Thus, $(a, b) \neq (b, a)$.

15.3 Mathematical Relation

Relations also exist outside human families. A **mathematical** or **number relation** is a relation that exists between sets of numbers.

 Example

A Mathematical relation "is a factor of" is defined from a set $X = \{2, 3, 4\}$ to another set $Y = \{1, 2, 3, 4, 5, 6, 7, 8\}$. Illustrate this relation diagrammatically.

Solution

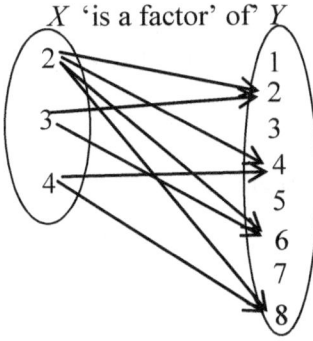

Notice that some elements in the codomain Y are not elements of any element in the domain. The **range** is the set of elements of the codomain, which are images of elements of the domain.
The illustrations above lead us to the definition of a relation.

A **relation** R is a rule that assigns an element $x \in A$ to another element $y \in A$ (relations in a set) or an element $x \in A$ to another element $y \in B$ (relations from one set to another). Therefore, a relation is a set of ordered pairs. Hence, we may write the relation "is a factor of" above defined from set $X = \{2,3,4\}$ to set $Y = \{1,2,3,4,5,6,7,8\}$ as $\Re = \{(2,2), (2,4), (2,6), (2,8). (3,3), (3,6), (4,4), (4,8)\}$.
Comparing $\Re$ with the Cartesian product $X \times Y$, we can see that a relation is a subset of a Cartesian product.

15.4 Notation

The statement 'a relates b' is denoted by $a\Re b$ or (a, b).

15.5 Ways of Defining Relations

We may define a relation in any of the following ways.
(i) Rule definition Method (ii) Formula Method
(iii) Ordered pair Method (iv) Table of values Method
(v) Graphical Method (vi) Arrow or pappy diagrams Method
(vii) Set builder notation Method

Example

Let $\Re$ be the relation "is half of" defined from the set $A = \{1, 2, 3, 4\}$ to $B = \{2, 4, 6, 8\}$. Redefine the relation using,
(i) Formula Method (ii) Arrow or pappy diagrams Method
(iii) Set builder notation Method (iv) Ordered pair Method
(v) Table of values Method (vi) Graphical Method

Solution

(i) Using the formula Method we can redefine R as $x = \dfrac{1}{2}y, x \in A, y \in B$

(ii) Using an arrow or pappy diagram we can represent the relation by,

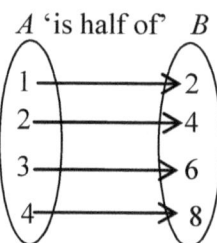

A 'is half of' B

(iii) By set builder notation method we can define the relation as

$$\Re = \left\{(x, y) : x = \dfrac{1}{2}y, x \in A, y \in B\right\}$$

(iv) As ordered pairs, the relation is $\Re = \{(1,2), (2,4), (3,6), (4,8)\}$

(v) Using a table of values the relation is

x	1	2	3	4
y	2	4	6	8

(vi) On a graph we can represent the relation as

Module 14, Topic 15: Relations and Functions

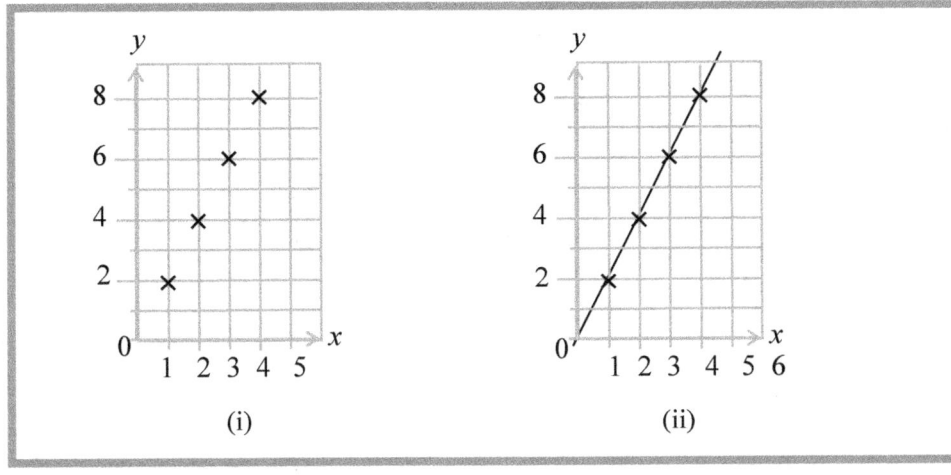

Notice that the graph in Figure (i) above is made up of discrete points. If the relation was defined on the set $\mathbb{R}$ of real numbers, a line as in Figure (ii) above will then connect these points. This type of relation is known as a **linear relation**. A linear relation is usually expressed as a linear equation of the form $y = mx + c$, where m and c are constants. Other forms of expressing a linear relation as an equation are treated in Topic 19.

 Example

1. A relation is defined on the set $X = \{a, b, c, d, e\}$ of children of the same family, as 'is a brother of'. Given that b and d are boys and a, c and e are girls. Draw an arrow diagram to illustrate this relation.

   Solution
   Since this relation is defined in a set, the representation will be as in the figure below.

   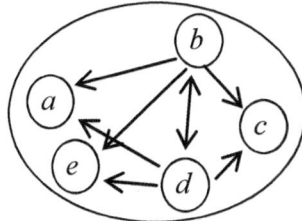

2. A relation is defined on that set $X = \{1,2,3,4,5,6\}$ as 'is a factor of'. Represent this relation on an arrow diagram.

Solution

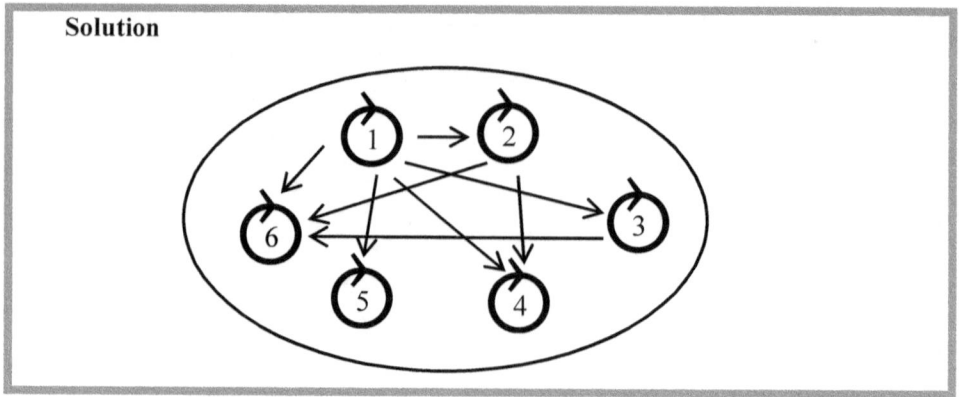

The above examples are examples of relations in a set. The loops ⟲ round each of the numbers is to indicate that each number is a factor of itself.

15.6 Inverse Relation

 Example

1. Consider the relation 'is a wife of' from the set $W = \{a, b, c, d\}$ of women to a set $M = \{w, x, y, z\}$ of men represented by the following arrow diagram. Define a relation from the set M to the set W and represent this relation on an arrow diagram.

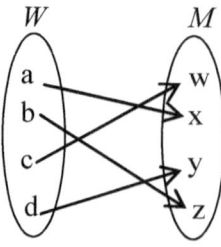

Solution
Clearly, we can define the relation "is the husband of" from the set M to the set W and represent it on arrow diagrams as follows.

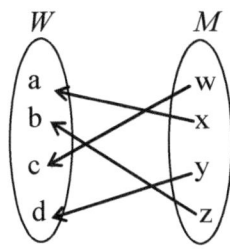 or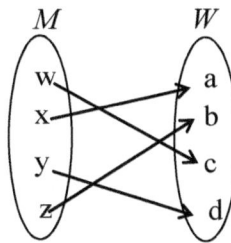

2. Find the inverse of each of the following relations
   (a) $\Re = \{(-4,4),(-2,2),(0,0),(2,-2),(4,-4)\}$
   (b)
x	0	1	2	3	4
y	0	2	4	6	8

   Solution
   (a) $\Re = \{(4,-4),(2,-2),(0,0),(-2,2),(-4,4)\}$
   (b)
x	0	2	4	6	8
y	0	1	2	3	4

3. Draw the graphs of the inverse relations in example 2.

   Solution

   (a)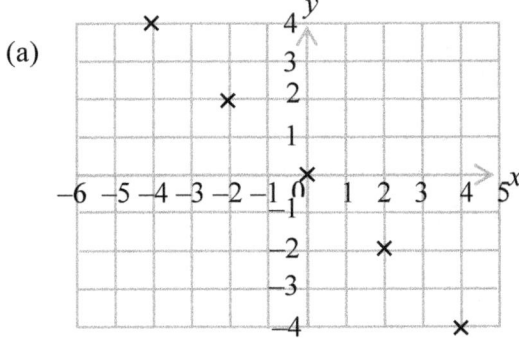

(b)

[graph showing plotted points on a coordinate grid with x-axis 0-9 and y-axis 0-4]

4. Bih who is the sister of Ambe is married to Anye and they are blessed with two children Manka and Neba. Ambe has a son called Fube.
   (i) Define a relation from the set of children of Bih to the son of Ambe
   (ii) Illustrate this relation on an arrow diagram.
   (iii) State the inverse of the relation in (i) and represent it on an arrow diagram.

Solution
The relation from the set of children of Bih to the son of Ambe is 'is a cousin of'.

A 'is a cousin of' B

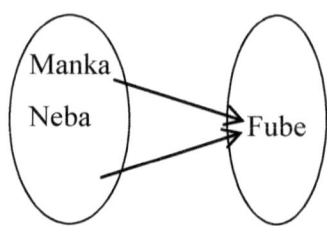

The inverse of the relation is 'is a cousin of'.

B 'is a cousin of' A

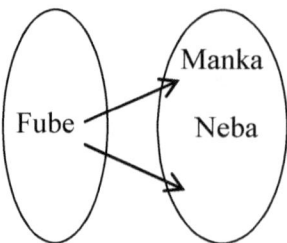

A relation such as 'is a cousin of' in example 4 above whose inverse is still 'is a cousin

Module 14, Topic 15: Relations and Functions

of' is said to be **self-inverse**.
Examples of mathematical relations, which are self-inverse, are,
(a) $\Re = \{(x, y): x + y = 8, x, y \in \mathbb{R}\}$
(b) $\Re = \{(-2, -2), (-1, -1), (0, 0), (1, 1)(2, 2)\}$

 Exercise 15:1

1. State the domain D and the range R of each of the following relations
   (a) $\{(0, 0), (1, 3), (2, 6), (3, 9)\}$ (b) $\{(0, 0), (1, 1), (1, -1), (4, 2), (4, -2)\}$
   (c) $\{(1, 15), (1, 20), (2, 20), (3, 25)\}$
2. Given the domain $\{1, 2, 3, 4\}$, determine the set of ordered pairs (x, y) of the relation defined by each of the following.
   (a) $y = x + 1$ (b) $y = 4 - x$
   (c) $y = 2(x - 3)$ (d) $y = x^2 + 1$
3. Given the domain $\{-4, -2, 0, 2, 4\}$, list the set of ordered pairs (x, y) of the relation defined by each of the following.
   (a) $\{(x, y): x = -y\}$ (b) $\{(x, y): y = 2x\}$
   (c) $\{(x, y): y = |x| - 2\}$ (d) $\{(x, y): y = x + 1\}$
4. If W is the set of all real numbers, represent each of the following relations graphically.
   (a) $\{(x, y): y = 9 - 2x, x, y \in W\}$ (b) $\{(x, y): y = 2x, x, y \in W\}$
   (c) $\{(x, y): x + y = 4, x, y \in W\}$ (d) $\{(x, y): y = x + 1, x, y \in W\}$
   (e) $\{(x, y): y = x, x, y \in W\}$ (f) $\{(x, y): x = y + 2, x, y \in W\}$
   (g) $\{(x, y): y = x^2, x, y \in W\}$
5. Draw a graph to represent the inverse of each of the relations in question 4 above.
6. Find the Cartesian product $A \times B$ of the sets $A = \{1, 3, 5\}$ and $B = \{2, 4, 6\}$.
7. Given that $A = \{-2, 1, 4\}$ and
   $B = \{10, 20\}$. List the ordered pairs of each of the following.
   (a) $A \times B$ (b) $B \times A$ (c) $A \times A$ (d) $B \times B$
8. Given the domain $\{-1, 1, 3, 5\}$, draw arrow or pappy diagrams for the relations defined by each of the following formulae.
   (a) $y = x + 1$ (b) $y = 4 - x$ (c) $y = 2(x - 3)$ (d) $y = x^2 + 1$
9. State the inverse of each of the following relations:
   (a) 'is a wife of'   (b) 'is a multiple of'   (c) 'is the teacher of'
   (d) 'is five times'   (e) 'is two more than one-third of'
10. The following figure illustrates the relation, 'is a factor of' in the set of numbers represented by $\{u, v, w, x, z\}$ which are integers.
    (a) Explain why each number has a loop.

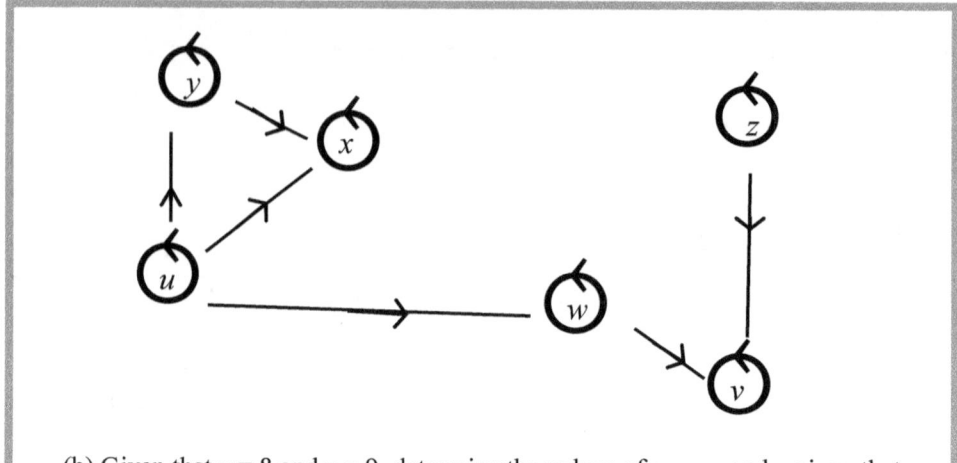

(b) Given that $y = 8$ and $z = 9$, determine the values of u, v, w and x given that u, v, w and x are as small as possible.

15.7 Properties of Relations in a Set

Reflexive

A relation $\Re$ in a set A is reflexive if every element in A relates to itself.

i.e. $\forall x \in A, x\Re x$

 Example

Let $E = \{1, 2, 3, 4\}$ and $\Re$ a relation in E defined by $x\Re x \Leftrightarrow x + x$ is even. Show that R is reflexive.

Solution

$1 + 1 = 2$ and 2 is even, $2 + 2 = 4$ and 4 is even, $3 + 3 = 6$ and 6 is even, $4 + 4 = 8$ and 8 is even.

$\therefore x + x = 2x$, and $2x$ is even. Hence, $x\Re x \quad \forall x \in E \Rightarrow$ relation is reflexive.

Symmetric

A relation $\Re$ in a set A, is symmetric if, x relates y implies y relates x for all x, y belonging to A.

i.e. $\forall x, y \in A, x\Re y \Rightarrow y\Re x$

Examples of symmetric relations are the relations 'is a friend of' and 'is a cousin of' defined over the set of people. For if, A is a friend to B then B is a friend to A and if X is a cousin to Y then Y is a cousin to X.

Example

Let $E = \{1, 2, 3, 4\}$ and $\Re$ a relation in E defined by $(x \Re y) \Leftrightarrow (x+y)$ is even. Show that R is symmetric.

Solution
$\Re = \{(1,1),(1,3),(2,2),(2,4),(3,1),(3,3),(4,2),(4,4)\}$. Observation of $\Re$ shows that $\forall x, y \in E, x \Re y \Rightarrow y \Re x$. Therefore, $\Re$ is symmetric.

Transitive

A relation $\Re$ in a set A is transitive if, x relates y and y relates z implies x relates z for all x, y, z belonging to A.
i.e. $\forall x, y, z \in A, x \Re y$ and $y \Re z \Rightarrow x \Re z$

Example

Show that the order relation 'is less than' in $\mathbb{R}$ is transitive.

Solution
Any 3 real numbers x, y and z are such that, if $x < y$ and $y < z$ then, $x < z$. For instance, $2 < 5$ and $5 < 8 \Rightarrow 2 < 8$.
Hence, the order relation 'is less than' in $\mathbb{R}$ is transitive.

Anti-symmetric

A relation $\Re$ in a set A is anti-symmetric if, x is related to y and y is related to x implies x is equal to y.

i.e. $\forall x, y \in A, x \Re y$ and $y \Re x \Rightarrow x = y$

Example

Let P be a set containing many sets and $\Re$ be the inclusion relation $\subseteq$. Show that the relation $\Re$ is anti-symmetric.

Solution

$A \subseteq B$ and $B \subseteq A \Rightarrow A = B$. Therefore, $\Re$ is anti-symmetric.

15.8 Equivalence relation

An **equivalence relation** is a relation $\Re$ in a set, which is reflexive, symmetric and transitive.

 Example

Let T be the set of all triangles and $\Re$ the relation 'is similar to'. Show that $\Re$ is an equivalence relation.

Solution
(i) Every triangle is similar to itself. $a\Re a \Leftrightarrow \Re$ is reflexive.
(ii) If a $\triangle ABC$ is similar to $\triangle XYZ$, then $\triangle XYZ$ must be similar to $\triangle ABC$.
   Therefore, $a\Re b \Leftrightarrow b\Re a$.
   Hence $\Re$ is symmetric.
(iii) If a $\triangle ABC$ is similar to $\triangle PQR$ and $\triangle PQR$ is similar to $\triangle XYZ$ then $\triangle ABC$ must be similar to $\triangle XYZ$. Therefore, $a\Re b$ and $bc \Rightarrow a\Re c$ and $\Re$ is transitive.
   Since the relation $\Re$ on T is reflexive, symmetric and transitive, it means $\Re$ is an equivalence relation.

 Exercise 15:2

1. The relation 'is less than or equal to' is defined on the set $A = \{1,2,3,4\}$.
   (i) Draw an arrow diagram to show the above relation on the set A.
   (ii) Explain why the relation is not an equivalence relation on the set A.
2. (i) Determine which of the properties: reflexive, symmetric, transitive, anti-symmetric apply to each of the following relations.
   (a) 'a divides b'; $a, b \in \mathbb{Z}$
   (b) $a \leq b; a, b \in \mathbb{Z}$
   (c) $|x| \leq |y|; x, y \in \mathbb{R}$
   (d) $x - y \leq 0; x, y \in \mathbb{R}$
   (e) $x + y \leq 1; x, y \in \mathbb{R}$
   (f) $x^2 = y^2; x, y \in \mathbb{R}$
   (g) $x - y$ is a multiple of 2π; $x, y \in \mathbb{R}$
   (h) $x^2 + y^2 = 1; x, y \in \mathbb{R}$
   (ii) State which of the relations in (i) are equivalence relations.

15.9 Order Relation

An **order relation** $\Re$ in a set A is a relation in the set, which is reflexive, transitive and anti-symmetric.

 Example

Define a relation $\leq$ in the set $\mathbb{R}$ of real numbers as 'is less than or equal to'. Show that $\Re$ is an order relation.

Solution

(i) $\forall a \in \mathbb{R}, a \leq a \Rightarrow \Re$ is reflexive.
(ii) $\forall x, y, z \in \mathbb{R}$ if $x \leq y$ and $y \leq z$ then, $x \leq z$. For instance, $2 \leq 5, 5 \leq 8$ and $2 \leq 8$ hence, the relation '$\leq$' defined in $\mathbb{R}$ is transitive.
(iii) $\forall x, y \in A$, $x\Re y$ and $y\Re x \Rightarrow x = y$. Hence, the relation '$\leq$' defined in $\mathbb{R}$ is anti-symmetric.
Since the relation '$\leq$' defined in $\mathbb{R}$ is reflexive, transitive and anti-symmetric then the relation is an order relation.

MAPPINGS AND FUNCTIONS

15.10 The Idea of a Function or a Mapping

Though the words function and mapping are interchangeable, there is a slight difference in their meanings. A function is necessarily a mapping but some mappings may not be functions. A function (sometimes for emphases, referred to as a "numerical function") usually refers to a mapping which involves the set of numbers. A mapping on the other hand may relate elements, which are neither numbers nor pro-numerals. Therefore, the set of all functions is a subset of the set of mappings.

Functions or mappings are special kinds of relations in which every element, x of the domain A relates to one and only one element y of the codomain B. To simplify, a function or mapping is a relation in which one and only one arrow leaves each element in the domain. A mapping is analogous to a relation 'is married to' defined from a set W of women to a set M of men in a society in which all women are married, polygamy is allowed but polyandry is forbidden. We should appreciate that this definition implies that for a mapping,

(i) Each element of the domain must map to one element of the codomain.
(ii) Not all elements of the codomain should necessarily be images.
(iii) An element of the codomain may be an image of more than one or even all

the elements of the domain.

In the arrow diagrams below, (i) is a mapping (function) from set A to set B but (ii) is not. If a, b, c, and d are real numbers then, (i) is a mapping but (ii) is not.

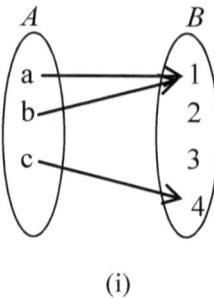 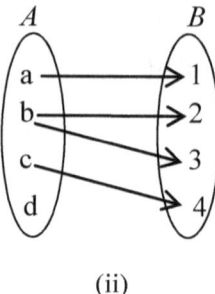

(i) (ii)

15.11 Function Notation

The two common function notations are:

(a) $f(x) = 5x + 4$ or $y = 5x + 4$ (b) $f: x \mapsto 5x + 4$

In the notation in (b), distinguish between the arrow $\mapsto$, used to map elements of two sets, and $\rightarrow$, used to map a set to another. For example $f: A \rightarrow B$.

15.12 Representation of Functions

Since a function is a relation it is usually represented using horizontal or vertical arrow diagrams or graphically by plotting the Cartesian coordinates.

 Example

1. Let $A = \{1, 2, 3, 4\}$ and $B = \{2, 4, 6, 8, 10\}$. Given the function $f: x \mapsto 2x$ defined from A to B. Represent this function using:
   (a) A horizontal arrow diagram,   (b) A Vertical arrow diagram,
   (c) Cartesian coordinates.

Solution

(a)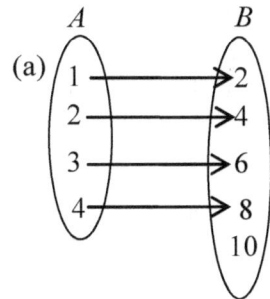

(b)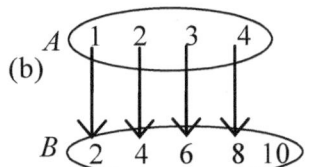

(c)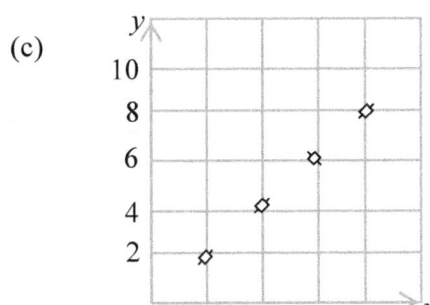

2. Which of the following relations are functions?
   (a) $\{(0,1),(0,2)\}$
   (b) $\{(1,0),(2,0)\}$
   (c) $\{(-3,1),(-3,2),(-3,3)\}$
   (d) $\{(1,-3),(2,-3),(3,-3)\}$
   (e) $\{(1,1),(\frac{1}{2},1),(\frac{1}{4},1),(-\frac{1}{4},1)\}$
   (f) $\{(-1,1),(-2,2),(-3,3),(2,-1),(3,-3),(1,-1)\}$

Solution
With the aid of arrow diagrams, the solution is clearer.

(a)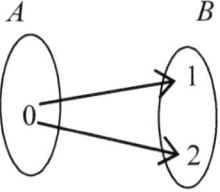

(b)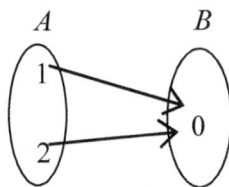

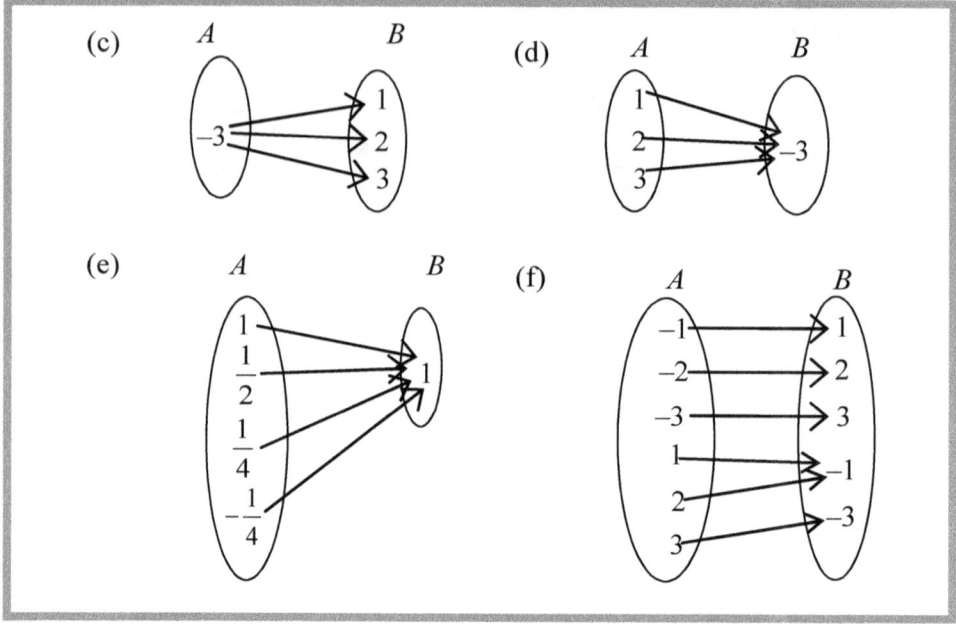

Note!
In arrow diagrams, the arrows of a mapping or function are parallel or convergent and never divergent.

From the arrow diagrams, we can see that in (b), (d), (e) and (f), the arrows are convergent. Hence, (b), (d), (e) and (f) are functions. (a) and (c) are not functions.

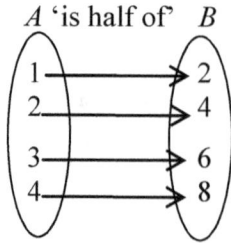

Alternatively, draw the graphs of the relations as shown in the following graphs. In this method if it is possible to draw a vertical straight line through any two points on the graph, then the relation is **not** a function.

Module 14, Topic 15: Relations and Functions

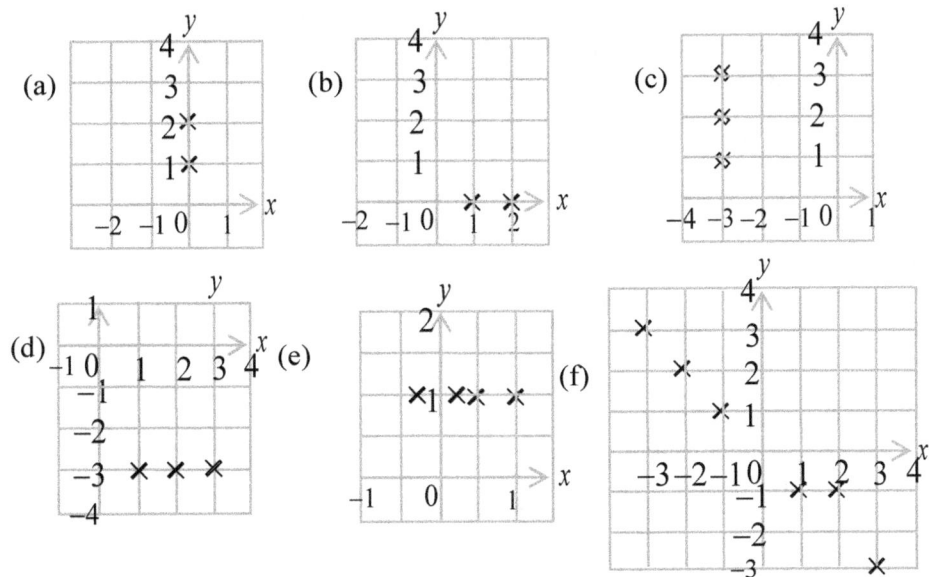

From the graphs above, we can see that in (b), (d), (e) and (f), we cannot draw a vertical line to pass through any two of the points. On the other hand in (a) and (c) we can draw a vertical line to pass through two or more points. Hence, (b), (d), (e) and (f) are functions while (a) and (c) are not.

Exercise 15:3

1. State which of the relations in the figures in (a) to (i) are mappings.

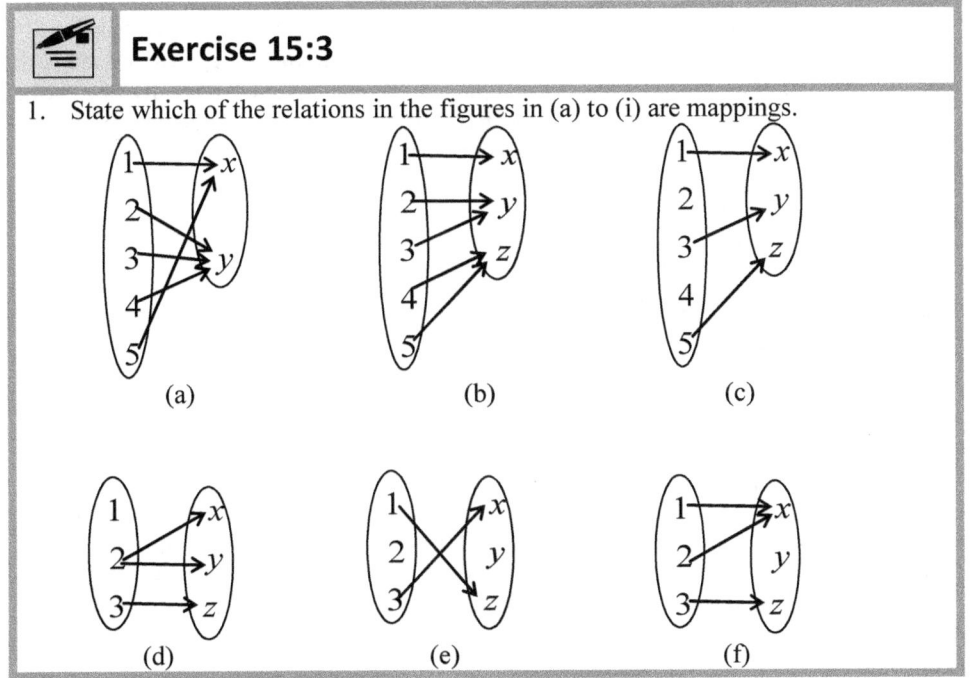

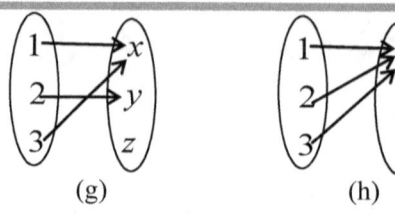
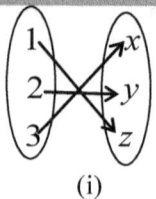

(g)   (h)   (i)

2. Let $A = \{x, y, z\}$ and $B = \{0, 1\}$.
    (a) Draw arrow diagrams of all the possible mappings from A into B.
    (b) How many mappings are there altogether?
3. Let $X = \{-2, -1, 0, 1, 2\}$ and given the function $f: X \longrightarrow \mathbb{Z}$ defined as $f: x \longmapsto 2x - 3$.
    (i) Represent this function using
        (a) A horizontal arrow diagram,   (b) A Vertical arrow diagram,
        (c) Cartesian coordinates.
    (ii) State the domain and range of the function.

15.13 Types of Mapping

We can classify mappings as one-one, onto, into or many-one mappings.

One-one mapping

A "one-one" mapping is a mapping in which each element in the domain A, has one element in the codomain B *and* vice versa. For instance, the function f, which maps every country to its capital city, is a one-one mapping, since each country has only one capital and there is no city, which is the capital of two different countries. The mapping in the figure below is a simple example of a one-one mapping.

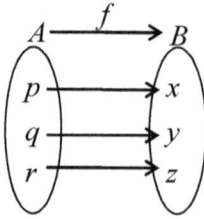

Onto mappings

An "**onto**" **mapping** is a mapping which is such that every element in the codomain is an image of at least one element in the domain. In other words an "onto" mapping is a mapping in which all the elements of the codomain are "used up". For instance; If $A = \{a, b, c, d\}$, $B = \{x, y, z\}$ and $f : A \longrightarrow B$ is defined as in Figure 23:18 then, f is an "onto" mapping.

Module 14, Topic 15: Relations and Functions

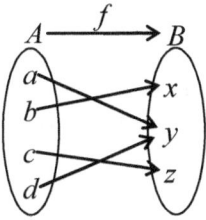

Into Mappings

An "into" mapping is a mapping, which is such that every element in the codomain is not an image of one element in the domain. In other words an "into" mapping is a mapping in which all the elements of the codomain are not "used up". For instance; If $A = \{a, b, c, d\}$, $B = \{x, y, z\}$ and $f : A \to B$ is defined as in the following figure, then, f is an "into" mapping.

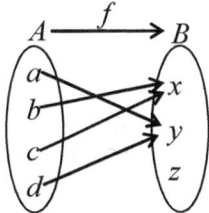

Many-one mapping

If a mapping is such that two or more elements in the domain give rise to the same image in the codomain, we say such a mapping is a many-one mapping. For instance, the function $f : x \mapsto x^2$ defined in the set $\mathbb{Z}$ of integers is a many-one mapping because each element in the codomain is the image of two elements in the domain.

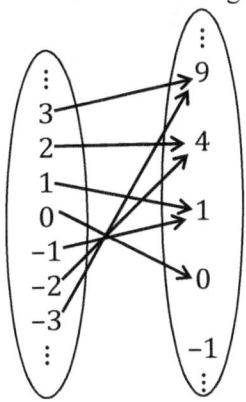

We can classify mappings as injections, surjections or bijections.

Competency Based Mathematics for Secondary Schools. Book 3

Injections

An injection or injective mapping or function is a mapping, which is such that every element $x \in A$ has a unique image $y \in B$. In other words, an injection is a one-to-one mapping of two sets such that each element of each set corresponds to only one element of the other set. For instance;
Let $A = \{1,2,3,4\}$ and $B = \{1,2,3,4,5,6\}$. The function $f: A \longrightarrow B$, defined by $f: x \longmapsto x + 1$, shown in the figure below is injective.

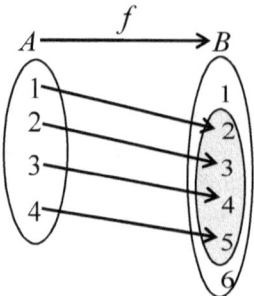

In figure 23:22, the range is shaded.

Notice that some elements in B may not be images of elements in A. Therefore, the range R is a proper subset of the codomain B. Hence for an injection, the cardinality of the domain A is always less than or equal to that of the codomain B.

$$n(A) \leq n(B)$$

Surjections

A surjection or surjective mapping or function is a function, which is such that every element $y \in B$ is the image of at least some element $x \in A$. In other words, all elements of the codomain B are "used up". For a surjection, there need not be a one-one correspondence between the elements of the domain and those of the range.
Hence for a surjection, the cardinality of the codomain B is always less than or equal to that of the domain A.

$$n(B) \leq n(A)$$

For instance;

Let $A = \{-3, -2, -1, 0, 1, 2, 3\}$ and $B = \{0, 1, 4, 9\}$. The relation $f: A \longrightarrow B$, defined by $f: x \longmapsto x$ shown in the figure below is surjective.

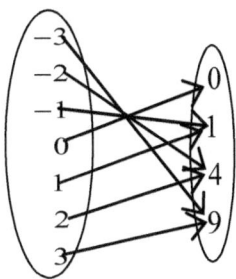

Let $f: \mathbb{N} \to E$, where $E = \{e, o\}$. Define a relation from $\mathbb{N}$ to E as follows,

$$f(n) = \begin{cases} o \text{ if } n \text{ is odd} \\ e \text{ if } n \text{ is even} \end{cases}$$

Since every element in E is an image of an element in $\mathbb{N}$, $f(n)$ is a surjection of $\mathbb{N}$ onto E.

Bijections

A bijection or bijective mapping or function is a function that is both injective (into) and surjective (onto). Therefore, a bijection is a one-one mapping which uses up all the elements in the codomain.

A mapping between two sets in which every element in each sets corresponds to only one element of the other sets for mapping in either direction.
For instance;

Let $A = \{1,2,3,4\}$ and $B = \{2,3,4,5\}$. The function $f: A \to B$ defined by $f: x \mapsto x + 1$, shown in the figure below is injective and surjective.

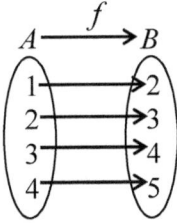

Exercise 15:4

1. Classify the functions in the following figure as
   (a) one-one   (b) many-one   (c) onto   (d) into.

   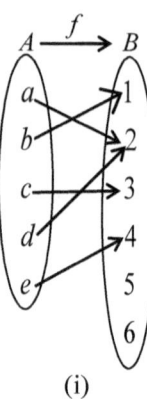
   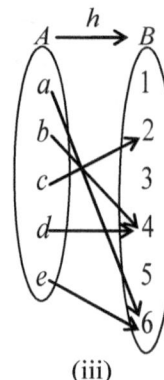

   (i)   (ii)   (iii)

2. State whether each of the following mappings is one-one, many-one.
   (i) Assign to each person in your village a number, which corresponds to his age.
   (ii) Assign to each city in your country, the corresponding number of people in the city.
   (iii) Assign to each book written by only one author its author.
   (iv) Assign to each country with a prime minister its prime minister.

3. In exercise 17:3 (3), state the number of mappings, which are:
   (i) one-one.   (ii) one-many.

4. Let $= A = \{1, 2, 3, 4, 5\}$. The figure below shows the function $f : A \rightarrow A$.
   (a) State the range of the function f.   (b) Is f an "onto" or "into" function?

   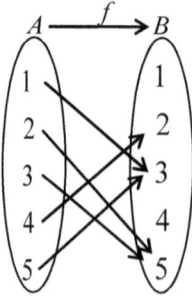

5. Classify the functions in Exercise 23:3 (2) as injections, surjections or bijections.
6. A correspondence $f: \mathbb{N} \rightarrow \mathbb{Z}$, is defined as follows. In each case determine whether f is a function, then classify the functions as injective or surjective.
   (a) $f(x) = \frac{1}{5}x + 3$ (b) $f(x) = x^2 - x$
   (c) $f(x) = x^2$ (d) $f(x) = \frac{1}{2}x^2 + 4$

Module 14, Topic 15: Relations and Functions

(e) $f(x) = x + 2$ (f) $f(x) = |x|$

7. Let $A = \{1, 2, 3, 4\}$, $B = \{w, x, y, z\}$ and $C = \{a, b, c\}$.

Draw an arrow diagram to represent:
(a) An injection, which is not a bijection.
(b) A surjection, which is not a bijection.
(c) A bijection.
(d) A function, which is not an injection, a surjection or a bijection.

15.14 Flow Charts

Functions are often analyzed using flow charts. More on flow charts have been treated in Topic 47. The reader may delay this section if necessary.

Example

Draw a flow chart to represent the function $f(x) \mapsto \dfrac{9-7x}{6}$

Solution

$x \to \boxed{\times(-7)} \xrightarrow{-7x} \boxed{+9} \xrightarrow{9-7x} \boxed{\div 6} \to \dfrac{9-7x}{6}$

Notice that in flow charts the order of operation is of utmost importance. For instance, in the function in the example above, it is imperative to first multiply x by -7 before adding 9 to the result.

15.15 Inverse Function

The inverse f^{-1} of a bijective function f is a function that performs the reverse process of what the function f does. For instance, if $g: x \mapsto x - 2$ then the inverse of g denoted by g^{-1} is $g^{-1}: x \mapsto x + 2$.

Note that only bijections have inverses

Finding the Inverse of a function

Suppose that $f: x \mapsto f(x)$. To find f^{-1}, equate $y = f(x)$ and solve for x, then substitute x for y to have $f^{-1}(x)$.

Competency Based Mathematics for Secondary Schools. Book 3

> **Example**
>
> Given that $f: x \mapsto \frac{2x-7}{5}$, find the inverse of f.
>
> **Solution**
>
> Let $\frac{2y-7}{5} = x$
>
> $\Rightarrow 2y - 7 = 5x$
>
> $2y = 5x + 7$
>
> $y = \frac{5x+7}{2}$
>
> $\therefore f^{-1}: x \mapsto \frac{5x+7}{2}$
>
> Alternatively, a flow chart of the function is drawn.
>
> $x \to \boxed{\times 2} \xrightarrow{2x} \boxed{-7} \xrightarrow{2x-7} \boxed{\div 5} \to \frac{2x-7}{5}$
>
> An inverse flow chart is drawn from the first. This is usually drawn in the opposite direction replacing each operation by its inverse.
>
> $\frac{5x+7}{2} \leftarrow \boxed{\div 2} \xleftarrow{5x+7} \boxed{+7} \xleftarrow{5x} \boxed{\times 5} \leftarrow x$
>
> Hence $f^{-1}: x \mapsto \frac{5x+7}{2}$

15.16 Composite Functions

A **composite function** $f \circ g$ or fg, is a function, which is made up of two simpler functions f and g. A composite function such as $f \circ g \circ h$ by convention is operated from right to left. The following figure shows $h = f \circ g(x) = 2x + 5$, the result of composing set A using the function $g: x \mapsto 2x$ followed by $f: x \mapsto x + 5$.

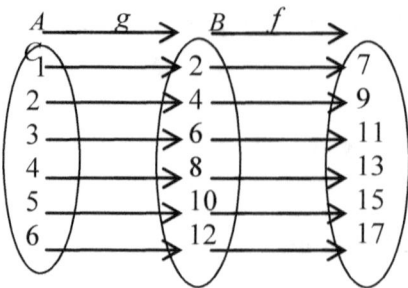

The composition of a function and its inverse is the **identity function**, usually denoted by $I: x \mapsto x$. The identity function always maps every element to itself.

Module 14, Topic 15: Relations and Functions

Example

1. Given that $f: x \mapsto 2x + 1$ and $g: x \mapsto x + 3x^2$, express $f \circ g$ in the form $f \circ g: x \mapsto \ldots$

   Solution
   $f \circ g(x) = f(3x^2) = 2(3x^2) + 1 = 6x^2 + 1$
   $\Rightarrow f \circ g: x \mapsto 6x^2 + 1$

2. The functions f and g are defined on $\mathbb{R}$, the set of real numbers, by
   $f: x \mapsto x + 4, g: x \mapsto \dfrac{1}{x+2}, x \neq -2$. Evaluate $f \circ g(2)$

   Solution
   $f \circ g(2) = f\left(\dfrac{1}{2+2}\right) = f\left(\dfrac{1}{4}\right) = \dfrac{1}{4} + 4 = \dfrac{17}{4}$

   Alternatively,
   $f \circ g(x) = f\left(\dfrac{1}{x+2}\right) = \dfrac{1}{x+2} + 4 = \dfrac{1 + 4x - 8}{x+2}$

   $f \circ g(2) = \dfrac{1 + 4(2) - 8}{2 + 2} = \dfrac{17}{4}$

15.17 Restricted Domain and Restricted Function

Consider the function $f: x \mapsto x^2$. The domain of definition of this function is $\mathbb{R}$ to $\mathbb{R}$. Clearly, but for the element 0 there will be two values of x for each image $f(x)$ as shown in the arrow diagram in figure (i) below.

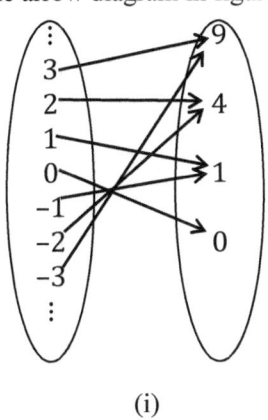
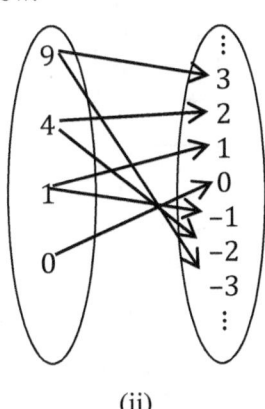

(i)   (ii)

The arrow diagram representing the inverse relation will be as shown in figure (ii) above. Notice that though the relation is a function, the inverse relation is not a function. Hence, the function $f: \mathbb{R} \to \mathbb{R}$; $f: x \mapsto x^2$ has no inverse. Suppose the function is redefined as $f: \mathbb{R}^+ \to \mathbb{R}$; $f: x \mapsto x^2$ the arrow diagram for the relation and the inverse relation will be as shown in figure (i) and (ii) below.

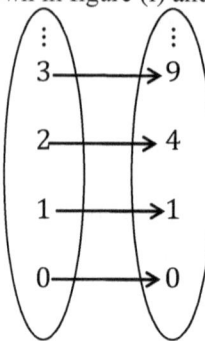
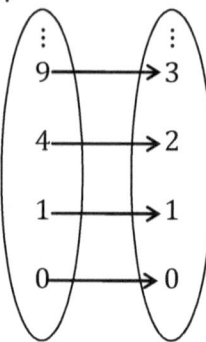

Clearly, both the relation and the inverse relation are functions. We call the new domain $\mathbb{R}+= [0, +\infty)$ of the function f the **restricted domain** of the function. A restricted domain is therefore, a domain that is smaller than the domain of definition of the function. We commonly use restricted domains to specify a one-to-one section of a function. We call a function with a restricted domain a **restricted function**.

Another example is the function $f: x \mapsto \frac{1}{x}$. The set $\mathbb{R}$ of real numbers cannot be the domain of this function, since there is no value for $f(0)$. Therefore, the domain of this function is the restricted domain $\mathbb{R} - \{0\}$.

 Exercise 15:5

1. The following diagram represents a mapping from set D to set C.

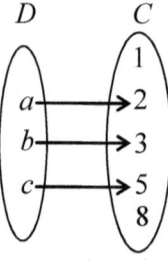

State (a) the element whose image is 2. (b) the range set.

2. The functions f and g, are defined in $\mathbb{R}$ such that
$$f(x) = 2x+1 \text{ and } g(x) = \frac{1}{x+2}, x \neq -2.$$ Find,

(i) The function $f^{-1}(x)$ that is the inverse of f. (ii) The value of $f \circ g(4)$.

Module 14, Topic 15: Relations and Functions

3. Given the functions $f: x \mapsto 3x-1, x \in \mathbb{R}$ and $g: x \mapsto x^2-1, x \in \mathbb{R}$.
   Solve the equation $f \circ g(x) = 0$.

4. f and g are functions defined on the set $\mathbb{R}$ of real numbers by
   $$f: x \mapsto 3x - 1 \text{ and } g: x \mapsto \frac{2x+1}{3}.$$
   Express in a similar manner (a) $g \circ f(x)$ (b) g^{-1}

5. Given the functions $f: s \mapsto 4s - 7$ and $g: s \mapsto s^2$ for $s \in \mathbb{R}^+$.
   Find (i) $f(2)$ (ii) $g^{-1} \circ f(4)$.

6. The function f is defined on the set $\mathbb{R}$ of real numbers, by
   $f: x \mapsto (x+1)(x+2)$.
   (i) $f(-3)$ (ii) Solve the equation $f(x) = 0$.

7. The function g is defined from a subset of $\mathbb{R}$ to a subset of $\mathbb{R}$ by $g(x) = x^2 - 3x$.
   (a) Find the range of g when the domain is $A = \{-2, 0, 2\}$.
   (b) Find the domain of g when the range is $B = \{0, -2\}$.

8. The function f is defined in the set of real numbers $\mathbb{R}$ by $f: x \mapsto 4 - x^2$.
   (a) $f(3)$ (b) Solve the equation $f(x) = 0$.

9. The functions f, g, and h are defined on $\mathbb{R}$ the set of real numbers as $f: x \mapsto x^2 - 3$, $g: x \mapsto x + 2$, $h: x \mapsto px + q$, where p and q are constants.
   (a) Evaluate $f(-2)$.
   (b) Find the composite function $g \circ f(x) = \ldots$.
   (c) Express g^{-1} in the form $g^{-1}: x \mapsto \ldots$.
   (d) Given that $g \circ f(x) = 3x + 1$, determine the values of p and q.

10. The functions f, g, and h are defined on $\mathbb{R}$ the set of real numbers as follows:
    $f: x \mapsto x + 2$, $g: x \mapsto x^2 - 1$, $h: x \mapsto 7x^2 - x + 2$.
    (a) Find a similar expression for f^{-1}.
    (b) Evaluate $g(\sqrt{2})$ and $h \circ f(0)$.
    (c) Find the value of x for which $h(x) = f(x)$.

11. The functions f, g and h are defined over the set $\mathbb{R}$ of real numbers by:
    $f: x \mapsto x^2 - 4$, $g: x \mapsto 3x - 6$, $h: x \mapsto (3x-8)(3x-4)$.
    (i) Prove that $f \circ g(x) = h$.   (ii) Calculate of $g \circ f(x)$.
    (iii) Express the quadratic equation $f \circ g(x) = g \circ f(x)$ in the form
    $ax^2 + bx + c = 0$.
    (iv) Solve the equation $f \circ g(x) - g \circ f(x) - 50 = 0$.

12. The functions f and g are defined on the set $\mathbb{R}$ of real numbers by
    $$f: x \mapsto \frac{x+3}{x-2} \text{ and } g: x \mapsto 2x - 1$$
    (i) Determine the domain and range of the function f.
    (ii) Find $g^{-1}(x)$. Hence or otherwise
    (iii) Find the function $h(x)$ such that $hg(x) = \frac{2x+1}{2(x-1)}$.

13. The functions p, q and r are defined in the set $\mathbb{R}$ of real numbers as follows:

$p: x \mapsto x^2 + 1$, $q: x \mapsto \frac{x}{3} + 1$, $r: x \mapsto 2x$

(a) Find expressions for $pr(x)$ and $rp(x)$.
(b) Solve correct to two decimal places, the equation $qp(x) = x + 1$.
(c) State the conditions necessary for p^{-1} to exist.
(d) Find expressions for the inverses p^{-1} and q^{-1} and express each one in a similar form to p, q.
(e) Explain why p^{-1} is not a function.

14. The functions f and g are defined in the set $\mathbb{R}$ of real numbers by
$f: x \mapsto 1 - x$ and $g: x \mapsto \frac{1}{2+x}, x \neq -2$

(a) Evaluate $g(-3)$ (b) Find $f^{-1}(x)$.
(c) Find to two significant figures the values of x for which
$f \circ g(x) = g \circ f(x)$.

15. In each of the following functions, the codomain is the set of real numbers. State the domain of each function.

(a) $f: x \mapsto \frac{\sqrt{x}}{x^2 - 9}$ (b) $f: x \mapsto \sqrt{x^2 - 16}$ (c) $f: x \mapsto 5x + 2$ (d) $f: x \mapsto \frac{1}{x}$ (e) $f: x \mapsto \frac{x-1}{(x+3)(x-2)}$ (f) $f: x \mapsto |2x|$

Multiple Choice Exercise 15

1. Given that $A = \{1,2,3,4,5\}$ and $B = \{-1,0,1,...,12\}$. A relation $\mathfrak{R}$ is defined from A to B as $a\,\mathfrak{R}\,b$ means $b = 3a - 4$. e.g. $1\,\mathfrak{R}\,_{-1}$ since, $-1 = 3(1) - 4$. The set of ordered pairs (a, b) of the relation, R is:
[A] $\{(1, -1),(2,2),(3,5),(4,8),(5,11)\}$ [B] $\{(-1,1),(2,2),(5,3),(8,4),(11,5)\}$
[C] $\{(2,2),(3,5)\}$ [D] $\{(2,2),(5,3)\}$

2. Given that $A = \{1,2,3,4,5\}$ and $B = \{-1,0,1,...,12\}$. A relation $\mathfrak{R}$ is defined from A to B as $a\,\mathfrak{R}\,b$ means $b = 3a - 4$. e.g. $1\,\mathfrak{R}\,_{-1}$ since, $-1 = 3(1) - 4$. It is true that:
[A] $\mathfrak{R}$ is an onto relation since, $\forall a \in A, b \in B, (a,b) \in A \times B$.
[B] $\mathfrak{R}$ is an onto relation since, $\forall a \in A, b \in B, (a,b) \notin A \times B$.
[C] $\mathfrak{R}$ is an onto relation since, $\forall a \in A, b \in B, (a,b) \supset A \times B$.
[D] $\mathfrak{R}$ is an onto relation since, $\forall a \in A, b \in B, (a,b) \subset A \times B$.

3. The properties, which both satisfy the relation 'is less than' are:
[A] reflexive and transitive.   [B] reflexive and symmetric.
[C] anti-symmetric and transitive   [D] symmetric and transitive.

4. The following figure shows a mapping involving two sets A and B described by the relation:
[A] $x \mapsto x$ [B] $x \mapsto x - 1$ [C] $x \mapsto x + 1$ [D] $x \mapsto 2x + 1$

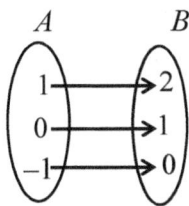

5. The ages of five children Cha, Abe, Eme, Bep and Dah, are such that Cha is older than Abe but younger than Eme. Eme is younger than Bep. Dah is older than Bep. The correct arrangement of the Children in the ascending order of their ages is:
   [A] Dah, Bep, Eme, Cha, Abe [B] Abe, Cha, Eme, Bep, Dah
   [C] Eme, Cha, Abe, Dah, Bep [D] Eme, Bep, Dah, Abe, Cha

6. A relation is defined on the set $X = \{1, 2, 3, 4, 5, 6\}$ as 'is a factor of'. The diagrams among the following which represent this relation is:

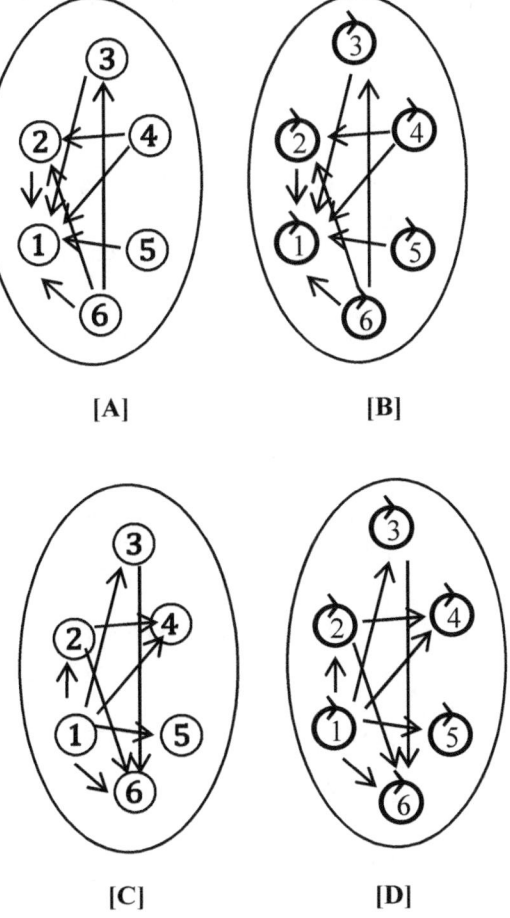

7. A relation $\Re$ is defined in a set $A = \{x, y, z\}$. $\Re$ is such that $x\Re y$ and $y\Re z \Rightarrow x\Re z$. Therefore:

[A] $\mathfrak{R}$ is reflexive.   [B] $\mathfrak{R}$ is symmetric.
[C] $\mathfrak{R}$ is transitive.   [D] $\mathfrak{R}$ is anti-Symmetric.

8. A relation $\mathfrak{R}$ is defined in a set $A = \{x, y, z\}$. If $\mathfrak{R}$ is said to be anti-symmetric. This means that:
   [A] $\forall x, y \in A,\ x\ \mathfrak{R}\ y \Rightarrow y\ \mathfrak{R}\ x$.
   [B] $\forall x, y, z \in A,\ x\ \mathfrak{R}\ y$ and $y\ \mathfrak{R}\ z \Rightarrow x\ \mathfrak{R}\ z$.
   [C] $\forall x, y \in A,\ x\ \mathfrak{R}\ y$ and $y\ \mathfrak{R}\ x \Rightarrow x = y$.
   [D] $\forall x \in A,\ x\ \mathfrak{R}\ x$.

9. A relation $\mathfrak{R}$ is defined in a set $A = \{x, y, z\}$. $\mathfrak{R}$ is such that, $x\ \mathfrak{R}\ x,\ \forall x \in A$. Therefore:
   [A] $\mathfrak{R}$ is reflexive.   [B] $\mathfrak{R}$ is symmetric.
   [C] $\mathfrak{R}$ is transitive.   [D] $\mathfrak{R}$ is anti-Symmetric.

10. A relation $\mathfrak{R}$ is defined in a set $A = \{x, y, z\}$. $\mathfrak{R}$ is symmetric means:
    [A] $\forall x, y \in A,\ x\ \mathfrak{R}\ y \Rightarrow y\ \mathfrak{R}\ x$.
    [B] $\forall x, y, z \in A,\ x\ \mathfrak{R}\ y$ and $y\ \mathfrak{R}\ z \Rightarrow x\ \mathfrak{R}\ z$.
    [C] $\forall x, y \in A,\ x\ \mathfrak{R}\ y$ and $y\ \mathfrak{R}\ x \Rightarrow x = y$.
    [D] $\forall x \in A,\ x\ \mathfrak{R}\ x$.

11. A relation $\mathfrak{R}$ is defined in a set $A = \{x, y, z\}$. $\mathfrak{R}$ is such that, $x\ \mathfrak{R}\ y \Rightarrow y\ \mathfrak{R}\ x,\ \forall x, y \in A$. Therefore, $\mathfrak{R}$ is:
    [A] $\mathfrak{R}$ is reflexive.   [B] $\mathfrak{R}$ is symmetric.
    [C] $\mathfrak{R}$ is transitive.   [D] $\mathfrak{R}$ is anti-Symmetric.

12. A relation $\mathfrak{R}$ is defined in a set $A = \{x, y, z\}$. $\mathfrak{R}$ is transitive means:
    [A] $\forall x, y \in A,\ x\ \mathfrak{R}\ y \Rightarrow y\ \mathfrak{R}\ x$.
    [B] $\forall x, y, z \in A,\ x\ \mathfrak{R}\ y$ and $y\ \mathfrak{R}\ z \Rightarrow x\ \mathfrak{R}\ z$.
    [C] $\forall x, y \in A,\ x\ \mathfrak{R}\ y$ and $y\ \mathfrak{R}\ x \Rightarrow x = y$.
    [D] $\forall x \in A,\ x\ \mathfrak{R}\ x$.

13. A relation $\mathfrak{R}$ is defined in a set $A = \{x, y, z\}$. $\mathfrak{R}$ is such that, $x\ \mathfrak{R}\ y$ and $y\mathfrak{R}x \Rightarrow x = y,\ \forall x, y \in A$. Therefore, $\mathfrak{R}$ is:
    [A] $\mathfrak{R}$ is reflexive.   [B] $\mathfrak{R}$ is symmetric.
    [C] $\mathfrak{R}$ is transitive.   [D] $\mathfrak{R}$ is anti-Symmetric.

14. A relation $\mathfrak{R}$ is defined in a set $A = \{x, y, z\}$. $\mathfrak{R}$ is reflexive means:
    [A] $\forall x, y \in A,\ x\ \mathfrak{R}\ y \Rightarrow y\ \mathfrak{R}\ x$.
    [B] $\forall x, y, z \in A,\ x\ \mathfrak{R}\ y$ and $y\ \mathfrak{R}\ z \Rightarrow x\ \mathfrak{R}\ z$.
    [C] $\forall x, y \in A,\ x\ \mathfrak{R}\ y$ and $y\ \mathfrak{R}\ x \Rightarrow x = y$.
    [D] $\forall x \in A,\ x\ \mathfrak{R}\ x$.

15. The diagram among the following which does not represent a function from set P to Q, is:

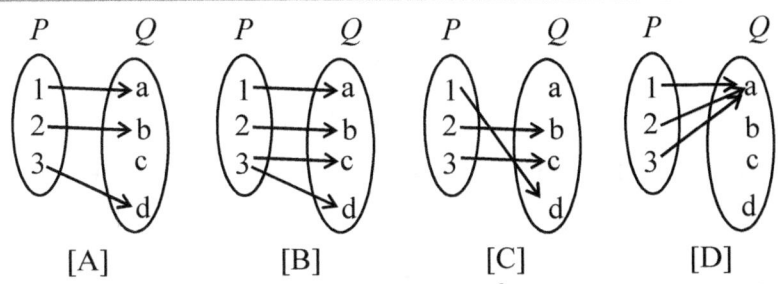

[A]   [B]   [C]   [D]

16. A function $f: \mathbb{R} \to \mathbb{R}$ is defined by $f: x \mapsto 2(x^2 + 1)$. The elements of the domain whose image is 10 are:
[A] −1 or 3   [B] 1 or 3   [C] −1 or −3   [D] −2 or 2

17. The function $f: \mathbb{R} \to \mathbb{R}$ is defined by $f: x \mapsto \begin{cases} -x \text{ when } x < -1 \\ 1 \text{ when } -1 \leq x \leq 1 \\ 2x - 1 \text{ when } x > 1 \end{cases}$

The graph of f in the following figure is:

[A]
[B]

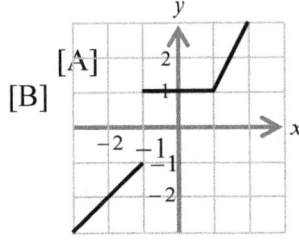 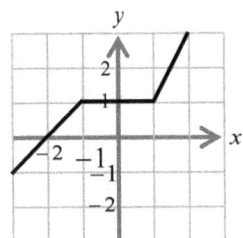

[C]
[D]

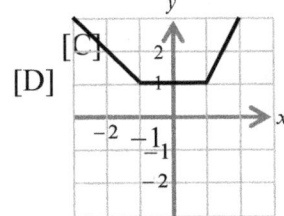 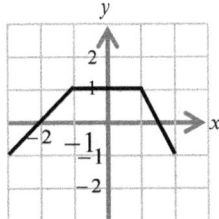

18. A function $f: \mathbb{R} \to \mathbb{R}$ is defined by $f: x \mapsto x + 2$. $f\left(\dfrac{1}{2}\right)$ is equal to:

[A] $\dfrac{5}{4}$ [B] 1   [C] $\dfrac{3}{2}$ [D] $\dfrac{5}{2}$

19. A function $f: \mathbb{R} \to \mathbb{R}$ is defined by $f: x \mapsto x + 2$. $f \circ f(8)$ is:
[A] 6   [B] 8   [C] 12   [D] 4

20. A function $f: \mathbb{R} \to \mathbb{R}$ is defined by $f: x \mapsto x + 2$. If $f\left(\dfrac{3}{a}\right) = f \circ f(a)$, the value of a must be:
[A] $-\dfrac{3}{2}$ [B] $\dfrac{3}{2}$ [C] 3   [D] −3

21. The function g is defined in $\mathbb{R}$, the set of real numbers, by $g: x \mapsto x + 4$. $g^{-1}: x$

is:

[A] $g^{-1}: x \mapsto \frac{4}{x}$ [B] $g^{-1}: x \mapsto x - 4$

[C] $g^{-1}: x \mapsto \frac{x}{4}$ [D] $g^{-1}: x \mapsto \frac{1}{x-4}$

22. The function f is defined in ℝ, the set of real numbers, by $f: x \mapsto \frac{1}{x+2}, x \neq -2$. $f(-4)$ is equal to:

[A] $-\frac{1}{2}$ [B] $\frac{1}{2}$ [C] $-\frac{1}{6}$ [D] $\frac{1}{6}$

23. The functions f and g are defined in ℝ, the set of real numbers, by $f: x \mapsto \frac{1}{x+2}, x \neq -2$ and $g: x \mapsto x + 4$. $g \circ f(2)$ is equal to:

[A] $\frac{3}{2}$ [B] $\frac{13}{2}$ [C] $\frac{17}{4}$ [D] $\frac{25}{6}$

24. Given the function f defined in the set ℝ of real numbers by $f: x \mapsto x^2 - 3x + 2$, $f(-3)$ is equal to:

[A] 2   [B] 20   [C] 7   [D] 5

25. Given the function f defined in the set ℝ of real numbers by $f: x \mapsto x^2 - 3x + 2$. The values of x for which $f(x) = 0$ are:

[A] 2 and 1   [B] −2 and −1   [C] 2 and −1   [D] −2 and 1

26. The functions f and g are defined in ℝ by $f: x \mapsto \sin x°$ and $g: x \mapsto 2x$. $f \circ g(45)$ is equal to:

[A] 0   [B] $\frac{\sqrt{2}}{2}$ [C] 1   [D] $\frac{1}{2}$

27. The functions f and g are defined in ℝ by $f: x \mapsto \sin x°$ and $g: x \mapsto 2x$. $g \circ f(30)$ is equal to:

[A] 0   [B] $\frac{\sqrt{2}}{2}$ [C] 1   [D] $\frac{1}{2}$

28. The function f is defined on ℤ, the set of integers, by $f: x \mapsto 1-2x$. $f(-4)$ is equal to:

[A] −7   [B] 7   [C] −9   [D] 9

29. The functions f and g are defined on ℤ, the set of integers, by $f: x \mapsto 1-2x$ and $g: x \mapsto 5x - k$. Given that $f \circ g(x) = g \circ f(x)$, the value of k is:

[A] −4   [B] 4   [C] $-\frac{4}{3}$ [D] $\frac{4}{3}$

30. The number of simple functions that make up the composite function $f: x \mapsto (2x + 5)^2$ is:

[A] 2   [B] 3   [C] 4   [D] 5

31. A function is defined on ℤ the set of integers as $f: x \mapsto 3 + x$. The image of −4 is:

[A] −1   [B] 1   [C] 4   [D] −4

32. The relation from set P to set Q in the arrow diagram below is:
    [A] 'is the square root of'   [B] 'is double'
    [C] 'is the square of'   [D] 'is 4 times'

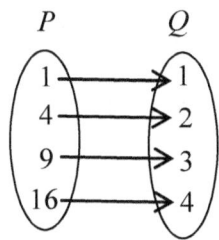

33. Given that $f: x \mapsto \dfrac{1}{x-2}$, then $f^{-1}: x \mapsto$:

    [A] $\dfrac{1}{x}-2$ [B] $\dfrac{1}{x}+2$ [C] $\dfrac{1}{x+2}$ [D] $\dfrac{1}{x-2}$

33. Given that $g(x) = x^2 - 6x, x \in \mathbb{R}$, then the value of m when $g(2m) = 10$ is:

    [A] $\sqrt{2}$ [B] 4   [C] −2 or 2   [D] 0 or 4

34.
x	−2	−1	0	1
y	−1	2	1	0

The correct graph and assertion about the relation in the above table is:

[A]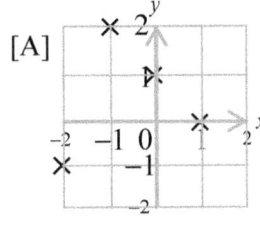
Not a function

[B]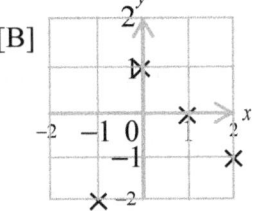
A function

[C]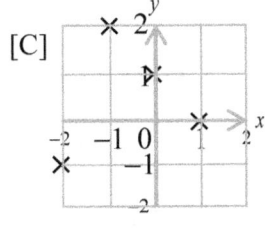
A function

[D]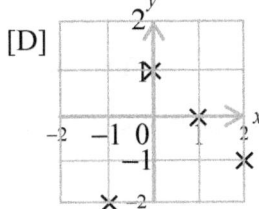
Not a function

35.
x	−1	0	1	1
y	1	−2	−1	2

The correct graph and assertion about the relation in the table above is:

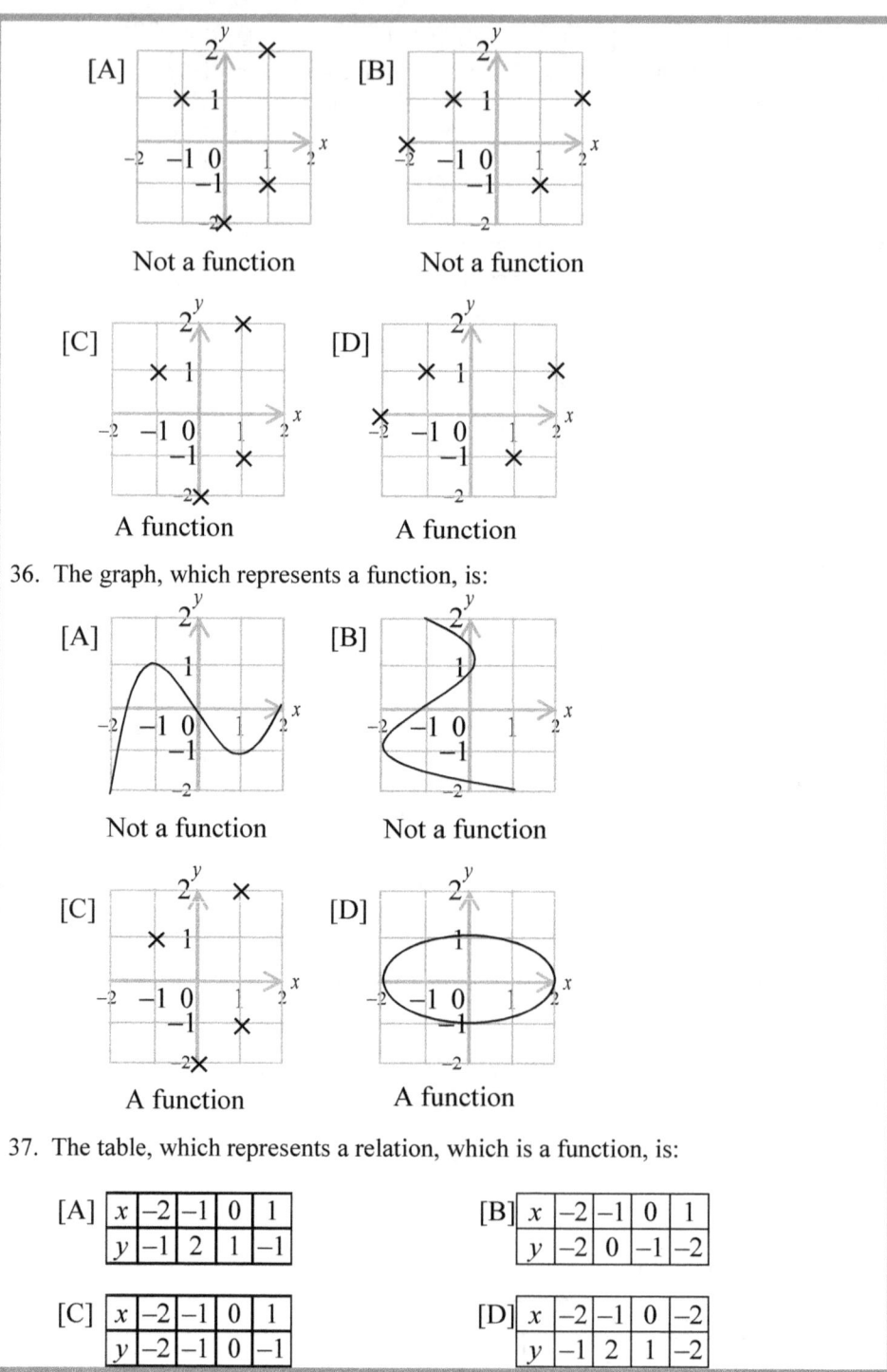

36. The graph, which represents a function, is:

37. The table, which represents a relation, which is a function, is:

[A]
x	−2	−1	0	1
y	−1	2	1	−1

[B]
x	−2	−1	0	1
y	−2	0	−1	−2

[C]
x	−2	−1	0	1
y	−2	−1	0	−1

[D]
x	−2	−1	0	−2
y	−1	2	1	−2

Answers to Structural Exercises

Exercise 1:1
1. (a) pen   (b) ink   (c) London   (d) i   (e) white paper
2. (a) $x \in A$ (b) $y \notin B$ (c) $n(G) = 3$
3. (a) Correct because b is an element of F
   (b) Not correct because {b} is a set and F is not a set of sets.
4. A is a set. A = {September, April, June, November}
   B is not a set. Beauty is relative.
   X is a set. We cannot list its members.
   G is not a set. Goodness is relative.
   P is a set. We cannot list its members.
   I is not a set. Not well defined.
   C is not a set. Members are of different families.
5. No. X is a set, but y is an element.
6. M = {2,4,6,8,10,12,14,16,18}
   F = M = {spades, hearts, diamonds, clubs}
   $V = \{a, e, i, o, u\}$
   A = {1,2,3,4,6,9,12,18,36}
   N = {2,3,4,5,6,7,8,9}
7. A = Odd numbers less than 14.
   B = Even numbers between 1 and 21.
   C = Fruits
   D = Adjectives
   E = Games
8. $X = \{x: x$ is a prime number less than 20$\}$
   $M = \{x: x$ is a factor of 42$\}$
   $C = \{x: x$ is a division in the North West Region of Cameroon$\}$
   $Y = \{x: x$ is a renowned river in Cameroon$\}$
   $T = \{x: x$ is a multiple of 3 less than 31$\}$

Exercise 1:2
1. (a) Trebleton   (b) singleton   (c) singleton   (d) Doubleton
   (e) doubleton   (f) Trebleton   (g) Doubleton   (h) none of the above
   (i) None of the above   (j) empty set   (k) Empty set
2. (a) 7   (b) 2   (c) 0   (d) 1   (e) 1
   (f) count and state the number of students in your class.
   (g) 8   (h) 5   (i) 14   (j) 20
3. (a) infinite   (b) finite   (c) finite   (d) Infinite

(e) finite   (f) infinite   (g) Finite   (h) finite   (i) finite   (j) Infinite
4. (a) ∉   (b) ∈   (c) ∈   (d) ∉   (e) ∈   (f) ∉

Exercise 1:3
1. (a) ⊂   (b) ∈   (c) ∉   (d) =   (e) ≠   (f) ≠
2. (a) False (b) False (c) False (d) True (e) False (f) True (g) False
   (h) True (i) True (j) False (k) False (l) False (m) False
3. { ∅ , {1},{2},{3},{1,2},{1,3},{2,3}, P}.   8 subsets
4. All are equal   5. {1,2,3,4,5}~{a, b, c, d, e}; {1,2,3,4,5} = {4,2,5,1,3} ;
   {2,4,6}~{Biology, Chemistry, Mathematics}   6. None
7. (a) False (b) True (c) False (d) False (e) False (f) False (g) True (h) True
8. (a) $a \in C$ (e) $e \notin A$ (f) $f \notin C$
9. $A = D$, $B = C$, $E = F$ $A \sim B$, $A \sim C$, $A \sim E$, $A \sim F$, $B \sim A$, $B \sim D$, $B \sim E$,
   $B \sim F$, $D \sim B$, $D \sim C$, $D \sim E$, $D \sim F$ $E \sim A$, $E \sim B$, $E \sim C$, $E \sim D$, $F \sim A$, $F \sim B$,
   $F \sim C$, $F \sim D$

Exercise 1:4
1. (i) A∪B   (ii) A∩B'   (iii) (A∪B)'   (iv) A∪B'   (v) (A∩B)∪(A∪B)'

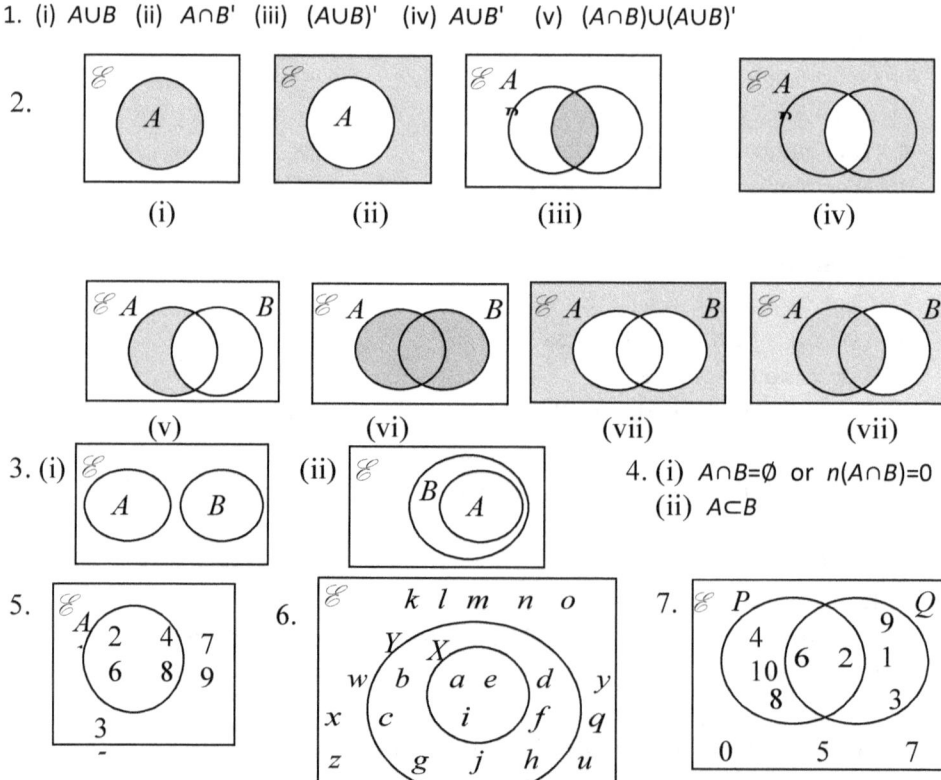

4. (i) A∩B=∅ or n(A∩B)=0
   (ii) A⊂B

Answers to Structural Exercises

8. 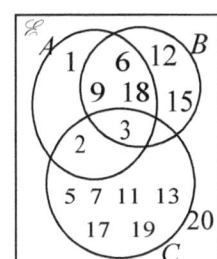   (a) {3}   (b) {3,6,9,18}

   (c) {3}   (d) {2,3}

9. (a) 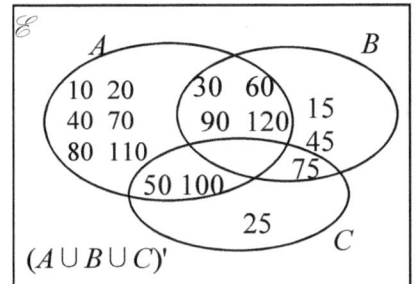   (b) n(A∪B) =16, n(A∩B∩C)= 0

Exercise 1:5

1. (a) 29 (b) 10 (c) 38 (d) 35

2. $n(N \cap B \cap M)' = 4$

3. (a) 37 (b) 7 (c) 62

4. (a) 11   (b) 26

Exercise 1:6

1. (a) The illegal sale of drugs is rising and the customs department is concerned with rising drug traffic.
   (b) The illegal sale of drugs is rising yet the customs department is concerned with rising drug traffic.
   (c) The customs department is not concerned with rising drug traffic.
   (d) The illegal sale of drugs is rising but the customs department is not concerned with rising drug traffic.
   (e) The illegal sale of drugs is not rising and the customs department is not concerned with rising drug traffic.
2. (a) $H' \cap S' \cap B$

(b)

[Venn diagram with three overlapping circles labeled H, S, B inside universal set ℰ, with the intersection region between H and S (outside B) shaded]

3. (a) $M \subset V$ (b) $(I \cap V) = 0$ or $I \cap V = \emptyset$
   (c) $n(M \cap H) \neq 0$ or $M \cap H \neq \emptyset$
4. (a) $P \cap T \cap N \neq \emptyset$ or $n(P \cap T \cap N) \neq 0$
   (b) $P \cap N \subset T$ (c) $P \cap N' \cap T' = 0$

Exercise 2:1

1. 216   2. 64   3. 72   4. 675   5. y^{10} 6. $8x^{10}$ 7. $60x^9$

8. a 9. $2a^3$ 10. $\dfrac{2x}{3y}$ 11. $\dfrac{2}{3b}$ 12. $\dfrac{1}{a^2}$ 13. $5x$ 14. $4ab^2$

15. $\dfrac{b}{a}$ 16. $-\dfrac{x}{3}$ 17. $2a^2$ 18. $5x$

Exercise 2:2

(i) 1. 4   2. $\dfrac{5}{4}$ 3. 5   4. $\dfrac{1}{36}$ 5. $\dfrac{3}{2}$ 6. 4   7. 2250   8. $\dfrac{8}{5}$ 9. $-\dfrac{31}{28}$ 10. $-\dfrac{17}{12}$ 11. $\dfrac{1}{12}$ 12. $\dfrac{1}{20}$ 13. 100   14. $25a^2$ 15. 3   16. $\dfrac{25}{3}$

(ii) (1) $x = \dfrac{7}{4}$ (2) $x = -\dfrac{5}{4}$ (3) $x = -\dfrac{3}{5}$ (4) $x = \dfrac{1}{2}$ (5) $x = -6$

(6) $x = \dfrac{5}{3}$ (7) $x = 3$ (8) $x = 5$ (9) $x = -\dfrac{4}{5}$ (10) $x = 6$

(11) $y = -\dfrac{3}{2}$ (12) $x = \dfrac{7}{3}$ (13) $x = \dfrac{9}{5}, y = \dfrac{27}{5}$ (14) $x = 2, y = 1$

(15) $x = 1, y = -2$

Answers to Structural Exercises

Exercise 2:3
(1) n (2) 2 (3) -1 (4) 3 (5) $\dfrac{1}{3}$ (6) 4

(7) 5 (8) $\dfrac{3}{2}$ (9) $\dfrac{3}{2}$ (10) $\dfrac{1}{3}$ (11) $\dfrac{1}{2}$ (12) $\dfrac{3}{2}$

(13) $y = 3^x$ (14) $10^x = 3$ (15) $n = 1$ (16) $n = 10$ (17) $y = 1$

(18) $x = 16$ (19) $x = \dfrac{1}{10}$ (20) $x = 3$

Exercise 2:4
1. (a) 3 (b) 7 (c) 3 (d) 1 (e) 2 (f) 3 (g) 2 (h) 2
2. (a) 1.8060 (b) 0.3010 (c) 0.1505
3. (i) 0.3891 (ii) -0.1743 4. 2.130

Exercise 3:1
1. (a) 2×2 (b) 2×3 (c) 3×1 (d) 3×3 (e) 3×2 (f) 1×1
   (g) 3×3 (h) 1×3 (i) 3×2 (j) 3×3
2. (a) square matrix (b) rectangular matrix (c) column matrix
   (d) diagonal matrix (e) rectangular zero matrix (f) trivial
   (g) square matrix (h) row matrix (i) rectangular matrix (j) unit matrix

Exercise 3:2
1. (a) $x = -3, y = 4$ (b) $x = 1, y = 0$ (c) $x = 5, y = 4$
   (d) $x = 3, y = -2$ (e) $x = 0, y = 2$
2. (a) $x = -3, y = 4, z = \dfrac{2}{5}$ (b) $x = -2, y = 4, z = -2$
3. (a) $x = -3, y = -4$ (b) $x = -2, y = 5$ (c) $x = -2, y = -7$
   (d) $x = 0, y = 0$ (e) $x = 5, y = -3$ (f) $x = 4, y = 1$

Exercise 3:3
1. (a) $\begin{pmatrix} 2 \\ -2 \end{pmatrix}$ (b) $\begin{pmatrix} 9 & 1 \\ 3 & 8 \end{pmatrix}$ (c) $\begin{pmatrix} -16 \\ 8 \end{pmatrix}$ (d) $\begin{pmatrix} -1 & 5 \\ -1 & -4 \end{pmatrix}$

2. (a) $\begin{pmatrix} 1 & -1 & 9 \\ 1 & 2 & 6 \\ 1 & 1 & -14 \end{pmatrix}$ (b) $\begin{pmatrix} -1 & 1 & -9 \\ -1 & -2 & -6 \\ -1 & -1 & 14 \end{pmatrix}$

3. (a) $\begin{pmatrix} 1 & 3 & 2 \\ 2 & 5 & 3 \end{pmatrix}$ (b) $\begin{pmatrix} 1 & 3 & 2 \\ 2 & 5 & 3 \end{pmatrix}$ (c) $\begin{pmatrix} 1 & 1 & 4 \\ 3 & 2 & 7 \end{pmatrix}$ (d) $\begin{pmatrix} 1 & 1 & -2 \\ 2 & 1 & -3 \end{pmatrix}$

(e) $\begin{pmatrix} -1 & -1 & 2 \\ -2 & -1 & 3 \end{pmatrix}$ (f) $\begin{pmatrix} -1 & 1 & 0 \\ -3 & 2 & -1 \end{pmatrix}$ (g) $\begin{pmatrix} 2 & 3 & 4 \\ 5 & 5 & 7 \end{pmatrix}$ (h) $\begin{pmatrix} 2 & 3 & 4 \\ 5 & 5 & 7 \end{pmatrix}$

(i) $\begin{pmatrix} 0 & 1 & -4 \\ -1 & 1 & -7 \end{pmatrix}$ (j) $\begin{pmatrix} 2 & 1 & 0 \\ 5 & 1 & 1 \end{pmatrix}$

(k) Addition of matrices is commutative.
(l) Subtraction of matrices is not commutative.
(m) Addition of matrices is associative.
(n) Subtraction of matrices is not associative.

4. (a) $\begin{pmatrix} 0 & 0 \\ 0 & 0 \end{pmatrix}$ (b) $\begin{pmatrix} 0 & 0 \\ 0 & 0 \end{pmatrix}$ (c) **A** is the additive inverse of **B**.

5. (a) $\begin{pmatrix} 3 & -1 \\ 2 & 5 \end{pmatrix}$ (b) $\begin{pmatrix} 3 & -1 \\ 2 & 5 \end{pmatrix}$ (c) $\emptyset$ is the additive identity.

6. $a = 1, b = -1, c = -6, d = 5, e = -7$
7. (a) $w = 6, x = 2, y = 0, z = -4$ (b) $w = 4, x = 3, y = 1, z = 5$

Exercise 3:4

1. $\begin{pmatrix} 2 & 0 & \frac{1}{2} \\ -\frac{3}{2} & -\frac{1}{2} & 1 \\ 0 & -4 & -\frac{5}{2} \end{pmatrix}$ 2. $x = 3$ 3. $x = 3, y = -2$

4. (i) $x = -2, y = 55$ (ii) $x = 9, y = -7$ (iii) $a = 2, b = 11$

Exercise 3:5

1. $\begin{pmatrix} -8 & 9 \\ 0 & -6 \end{pmatrix}$ 2. $(-13 \quad -35 \quad 17)$

Answers to Structural Exercises

3. (i) $\begin{pmatrix} 3 & 7 \\ 5 & -2 \end{pmatrix}$ (ii) $\begin{pmatrix} 3 & 7 \\ 5 & -2 \end{pmatrix}$ (iii) $\begin{pmatrix} 44 & 7 \\ 5 & 39 \end{pmatrix}$

4. (a) $\begin{pmatrix} 1 & 3 \\ 5 & 4 \end{pmatrix}$ (b) $\begin{pmatrix} 1 & 3 \\ 5 & 4 \end{pmatrix}$ (c) $\begin{pmatrix} 0 & 0 \\ 0 & 0 \end{pmatrix}$

(d) $\begin{pmatrix} 0 & 0 \\ 0 & 0 \end{pmatrix}$ (e) $\begin{pmatrix} 1 & 2 \\ 5 & 4 \\ 2 & 6 \end{pmatrix}$ (f) $\begin{pmatrix} 0 & 0 \\ 0 & 0 \\ 0 & 0 \end{pmatrix}$

For any compatible matrices, multiplication by

(i) $\begin{pmatrix} 1 & 0 \\ 0 & 1 \end{pmatrix}$ leaves the matrix unchanged.

(ii) $\begin{pmatrix} 0 & 0 \\ 0 & 0 \end{pmatrix}$ gives the result $\begin{pmatrix} 0 & 0 \\ 0 & 0 \end{pmatrix}$.

5. (i) (a) $\begin{pmatrix} 0 & -20 \\ 5 & 5 \end{pmatrix}$ (b) $\begin{pmatrix} 7 & 18 \\ 6 & 19 \end{pmatrix}$ (c) $\begin{pmatrix} 0 & -10 \\ 5 & 15 \end{pmatrix}$ (d) $\begin{pmatrix} 7 & 1 \\ 6 & 8 \end{pmatrix}$ (e) $\begin{pmatrix} -3 & 13 \\ -4 & 14 \end{pmatrix}$

(f) $\begin{pmatrix} 10 & -30 \\ -15 & 55 \end{pmatrix}$ (g) $\begin{pmatrix} 10 & -30 \\ -15 & 55 \end{pmatrix}$ (h) $\begin{pmatrix} 4 & -1 \\ 2 & 7 \end{pmatrix}$ (i) $\begin{pmatrix} 0 & -7 \\ 0 & -1 \end{pmatrix}$ (j) $\begin{pmatrix} 14 & -11 \\ 22 & 47 \end{pmatrix}$

(k) $\begin{pmatrix} 0 & 7 \\ 0 & 1 \end{pmatrix}$ (l) $\begin{pmatrix} -7 & -38 \\ -1 & -4 \end{pmatrix}$ (m) $\begin{pmatrix} -14 & -49 \\ -2 & -7 \end{pmatrix}$ (n) $\begin{pmatrix} 7 & -22 \\ 21 & 54 \end{pmatrix}$ (o) $\begin{pmatrix} 7 & 18 \\ 1 & -6 \end{pmatrix}$

(ii) (a) **AB ≠ BA** (b) **(AB)C ≠ A(BC)** (c) $(A + B)^2 \neq A^2 + 2AB + B^2$

(d) $(A - B)^2 \neq A^2 - 2AB + B^2$ (e) $(A + B)(A - B) \neq A^2 - B^2$

6. (a) $\begin{pmatrix} 5x - 3y \\ -4x + y \end{pmatrix}$ (b) $\begin{pmatrix} 3a + 4b \\ 5a + b \end{pmatrix}$ (c) $\begin{pmatrix} 2r + s \\ -2r + 3s \end{pmatrix}$ (d) $\begin{pmatrix} 4p + q \\ 2p \end{pmatrix}$

(e) $\begin{pmatrix} 4x + 2y \\ 3x + 5y \end{pmatrix}$ (f) $\begin{pmatrix} -2x - 3y \\ -2x + y \end{pmatrix}$

Exercise 3:6

(a) $\begin{pmatrix} 1 & 0 \\ 3 & 4 \end{pmatrix}$ (b) $\begin{pmatrix} 1 \\ 2 \\ 7 \end{pmatrix}$ (c) $\begin{pmatrix} -1 & 0 & 1 \end{pmatrix}$ (d) $\begin{pmatrix} 2 & 1 & -3 \\ 0 & 8 & 5 \\ -4 & 6 & 5 \end{pmatrix}$

Exercise 3:7

1. -2 2. $\dfrac{10}{3}$ 3. 2 4. ± 2 5. $\begin{pmatrix} -2 & -2 \\ 2 & 2 \end{pmatrix}$

6. $\begin{pmatrix} 11 & -7 \\ -16 & 10 \end{pmatrix}, -2$ 7. -4 or 11 8. -1 or 3 9. 3 or 7

Exercise 4:1
1. $\triangle PTQ \equiv \triangle RTS$, $\triangle PQR \equiv \triangle SRQ$, ASA
2. $\triangle ABC \equiv \triangle AED$, $\triangle ABD \equiv \triangle AEC$, SAS
3. $\triangle XOY \equiv \triangle XOZ$, SAS
4. $\triangle POM \equiv \triangle QLN$, ASA
5. $\triangle POS \equiv \triangle ROQ$, SAS or ASA or SSS
   $\triangle POQ \equiv \triangle ROS$, ASA
   $\triangle PQS \equiv \triangle RSQ$, SAS
   $\triangle PRS \equiv \triangle RPQ$, SAS
6. $\triangle PTQ \equiv \triangle RTS$, ASA
7. $\triangle WOZ \equiv \triangle XOY$, ASA
   $\triangle WYZ \equiv \triangle XZY$, ASA
8. Two sets by SAS or ASA or SSS
   $\triangle ADF \equiv \triangle DBE \equiv \triangle FEC \equiv \triangle DFE$
   $\triangle DIG \equiv \triangle IFH \equiv \triangle IGH \equiv \triangle GHE$

Exercise 4:2
1. (i) 8 cm (ii) 30 cm 2. (b), (h) 3. (i) 14 cm (ii) 44.1 cm^2
4. 19.1 cm 5. $m = 3$, $n = 13.5$

Exercise 4:3
1. 25 km^2 2. 20 cm and 24 cm; 1:16 3. 3 cm and 5 cm; 1:9 4. 4
5. 9 cm^2 6. 20.88 cm^2 7. 2 8. (b) 8 cm (c) 16:1

Exercise 4:4
1. 250 cm^3 2. 268 cm^3, 35 cm 3. 4.5 4. 81 cm^3
5. 121.5 cm^3 6. 10.5 7. 64:15625 8. 2847 cm^3

Answers to Structural Exercises

Exercise 5:1

1.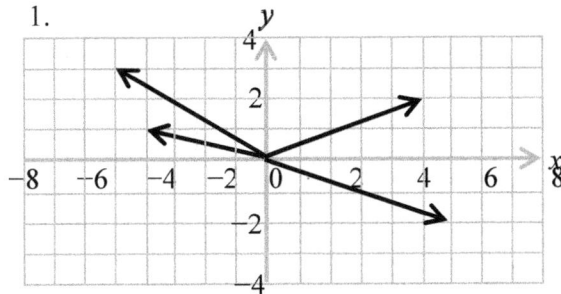

2. $AB = \begin{pmatrix} 3 \\ -11 \end{pmatrix}$, $CD = \begin{pmatrix} 7 \\ 6 \end{pmatrix}$, $EF = \begin{pmatrix} 0 \\ 8 \end{pmatrix}$, $GH = \begin{pmatrix} 0 \\ 8 \end{pmatrix}$

3.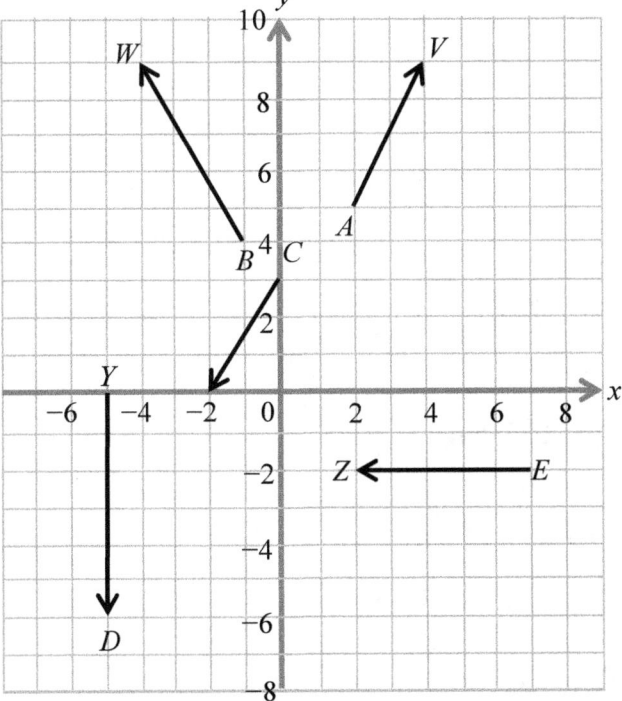

Competency Based Mathematics for Secondary Schools. Book 3

4.
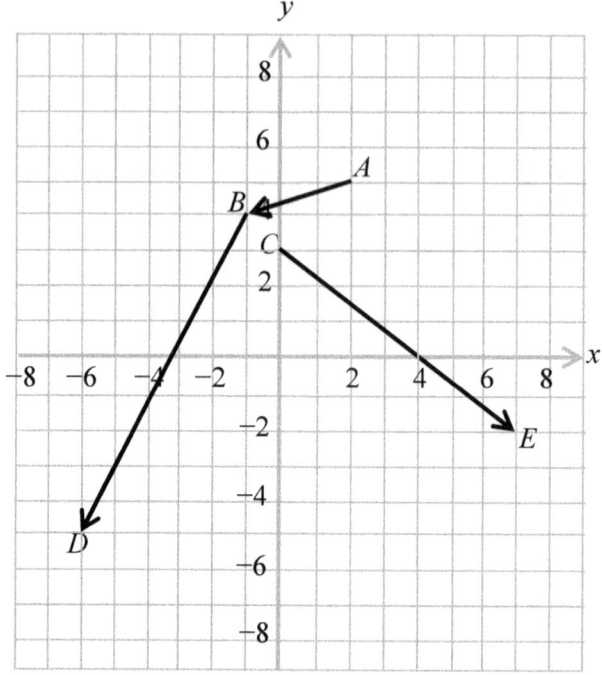

Exercise 5:2

1.
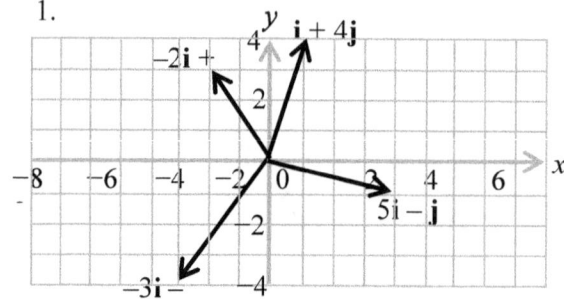

2. (i) $2i + 5j$ (ii) $-5i + 3j$ (iii) $-4i - j$ (iv) $6i - 2j$

3. (i) $3p + 5q$ (ii) $-7p + 2q$ (iii) $-4p - 2q$

4. (i) $\begin{pmatrix} -3 \\ 2 \end{pmatrix}$ (ii) $\begin{pmatrix} 2 \\ -5 \end{pmatrix}$ (iii) $\begin{pmatrix} 9 \\ 4 \end{pmatrix}$ (iv) $\begin{pmatrix} -7 \\ -3 \end{pmatrix}$

Answers to Structural Exercises

Exercise 5:3

1. (i) 13 (ii) $\sqrt{13}$ (iii) $\sqrt{58}$
   (iv) 13 (v) 10 (vi) 5
2. (i) $\sqrt{41}$ (ii) $4\sqrt{2}$ (iii) $\sqrt{13}$
3. $\mathbf{i}, \frac{3}{5}\mathbf{i}+\frac{4}{5}\mathbf{j}, \mathbf{j}, \frac{1}{\sqrt{2}}\mathbf{i}-\frac{1}{\sqrt{2}}\mathbf{j}, -\frac{\sqrt{3}}{2}\mathbf{i}+\frac{1}{2}\mathbf{j}$
4. 5
5. $\sqrt{13}$

Exercise 5:4

1. (a) $\mathbf{OA} = \begin{pmatrix} -10 \\ 12 \end{pmatrix}$, $\mathbf{OB} = \begin{pmatrix} 8 \\ 10 \end{pmatrix}$, $\mathbf{OC} = \begin{pmatrix} 6 \\ 6 \end{pmatrix}$,

   $\mathbf{OD} = \begin{pmatrix} 6 \\ -4 \end{pmatrix}$, $\mathbf{OE} = \begin{pmatrix} -6 \\ -2 \end{pmatrix}$, $\mathbf{OF} = \begin{pmatrix} -8 \\ 2 \end{pmatrix}$,

   (b) **OA** = −10i+12j, **OB** = 8i+10j, **OC** = 6i+6j, **OD** = 6i−4j, **OE** = −6i−2j, **OF** = −8i+2j

2. (i) (a) $\mathbf{OA} = \begin{pmatrix} -1 \\ 5 \end{pmatrix}$, $\mathbf{OB} = \begin{pmatrix} 4 \\ -7 \end{pmatrix}$, $\mathbf{OC} = \begin{pmatrix} -9 \\ 3 \end{pmatrix}$, $\mathbf{OD} = \begin{pmatrix} -3 \\ 6 \end{pmatrix}$

   (b) **OA** = −i + 5j, **OB** = 4i −7j, **OC** = −9i+3j, **OD** = −3i + 6j

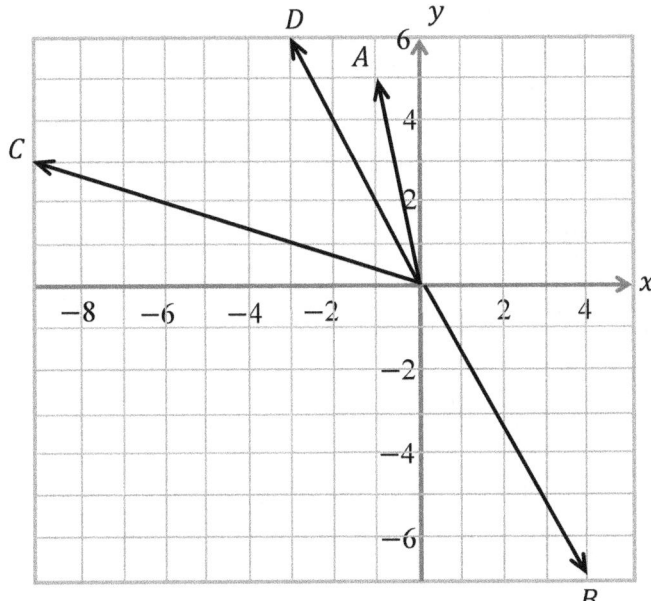

3. **AB = DC, EF = KL, ST = MN, PQ = XY.**

4. a and l, b, g, j and m, c and i, d and k, e and p.   5. **OD** and **OE**.

Exercise 5:5

1. (i)

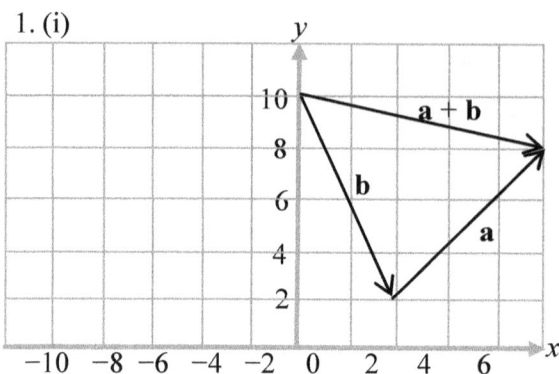

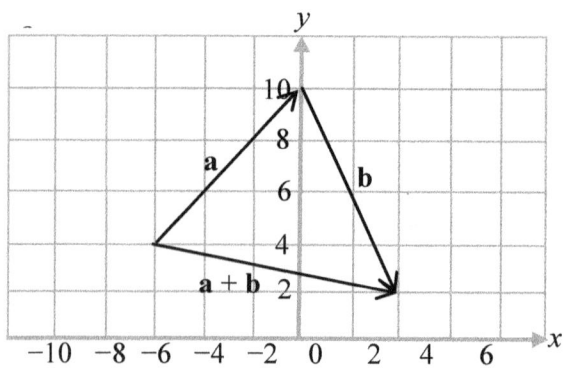

(ii)

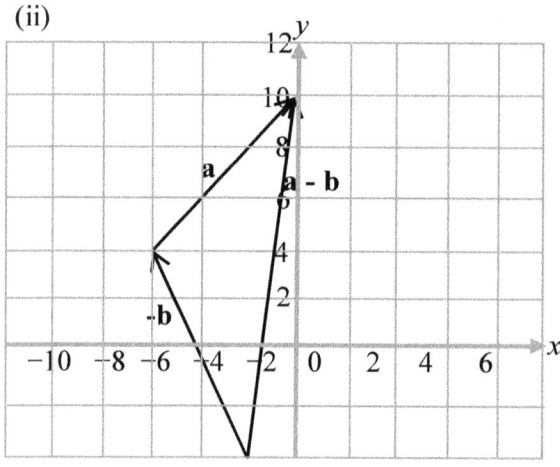

Or

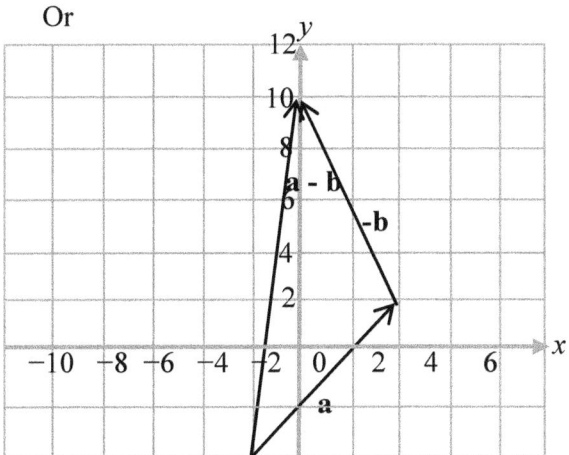

(iii)

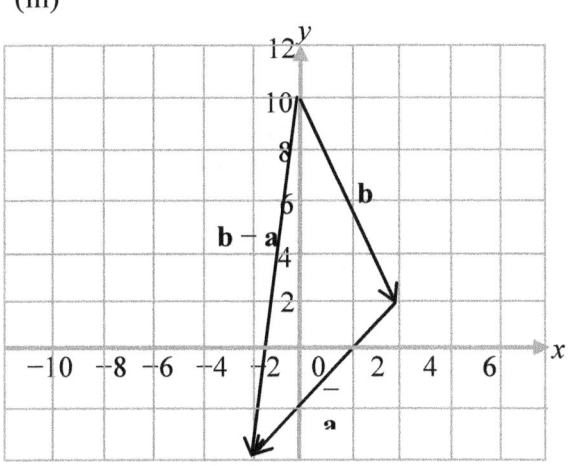

Or

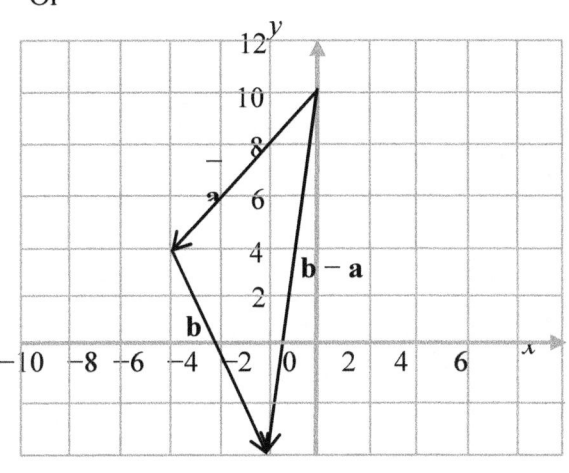

Competency Based Mathematics for Secondary Schools. Book 3

2. (i) −2i + 2j (ii) −2i + 2j (iii) 8i −4j (iv) −8i + 4j
3. (a) i+2j (b) -2i-5j (c) -2i+7j
   (d) i − 2j (e) $\begin{pmatrix} 2 \\ 4 \end{pmatrix}$ (f) $\begin{pmatrix} -7 \\ -5 \end{pmatrix}$ (g) $\begin{pmatrix} 3 \\ -2 \end{pmatrix}$ (h) $\begin{pmatrix} -8 \\ 3 \end{pmatrix}$

4. (i) $\begin{pmatrix} -1 \\ 7 \end{pmatrix}$ (ii) $\begin{pmatrix} -1 \\ 7 \end{pmatrix}$ (iii) $\begin{pmatrix} 5 \\ 3 \end{pmatrix}$ (iv) $\begin{pmatrix} -5 \\ -3 \end{pmatrix}$

Conclusions:
a + b = b + a and **a − b = −(b − a)**

Exercise 5:6
1. $a = 2$ and $b = -1$
2. (i) $\begin{pmatrix} 9 \\ 4 \end{pmatrix}$ (ii) $\begin{pmatrix} -1 \\ -10 \end{pmatrix}$ (iii) $\begin{pmatrix} 1 \\ 10 \end{pmatrix}$ (iv) $\begin{pmatrix} 3 \\ 8 \end{pmatrix}$ (v) $\begin{pmatrix} 3 \\ 8 \end{pmatrix}$ (vi) $\begin{pmatrix} 12 \\ -9 \end{pmatrix}$ (vii) $\begin{pmatrix} 8 \\ 5 \end{pmatrix}$

(viii) $\begin{pmatrix} -21 \\ 25 \end{pmatrix}$ (ix) $\begin{pmatrix} \frac{14}{3} \\ -\frac{11}{3} \end{pmatrix}$ (x) $\begin{pmatrix} 31 \\ -11 \end{pmatrix}$

3. $\begin{pmatrix} 3 \\ -3 \end{pmatrix}$, **x + y = 3z** i.e. **x + y** is a scalar multiple of **z**. 4. (a) Since 4**CD** = **AB**, **AB** is parallel to **CD**. (b) 4 : 1

5. $5\sqrt{2}$ 6. $3u + v = 5$, $u + v = 3$; $u = 1, v = 2$

Exercise 6:1
1. 10 cm   2. 4 cm   3. 41 cm   4. 18 cm   5. 12 cm and 16 cm
6. (a), (c), (e), (h)

Exercise 6:2
1. (a) $\frac{12}{5}$ (b) $\frac{12}{13}$ (c) $\frac{5}{13}$ (d) $\frac{5}{12}$ (e) $\frac{5}{13}$ (f) $\frac{12}{13}$

(ii) (a) $\frac{40}{9}$ (b) $\frac{40}{41}$ (c) $\frac{9}{41}$ (d) $\frac{9}{40}$ (e) $\frac{9}{41}$ (f) $\frac{40}{41}$

2. (a) $\frac{3}{4}$ (b) $\frac{4}{5}$ (c) $\frac{3}{5}$ (d) $\frac{15}{8}$ (e) $\frac{8}{17}$ (f) $\frac{15}{17}$

3. (a) $\frac{7}{25}$ (b) $\frac{24}{25}$ 4. (i) $\frac{\sqrt{3}}{2}$ (ii) $\frac{1}{2}$ (iii) $\sqrt{3}$

5. (i) $\frac{\sqrt{2}}{2}$ (ii) $\frac{\sqrt{2}}{2}$ (iii) 1

Answers to Structural Exercises

6. $\cos x = \dfrac{\sqrt{n^2-m^2}}{n}$, $\tan x = \dfrac{m}{\sqrt{n^2-m^2}} = \dfrac{m\sqrt{n^2-m^2}}{n^2-m^2}$ 7. (a) $\dfrac{3}{5}$ (b) $\dfrac{3}{4}$

8. (a) $\dfrac{4}{5}$ (b) $\dfrac{3}{5}$ (c) $\dfrac{3}{4}$ (d) $\dfrac{3}{5}$ (e) $\dfrac{4}{5}$ (f) $\dfrac{4}{3}$
 (g) $\dfrac{5}{13}$ (h) $\dfrac{12}{13}$ (i) $\dfrac{12}{5}$ (j) $\dfrac{12}{13}$ (k) $\dfrac{5}{13}$ (l) $\dfrac{5}{12}$

Exercise 6:3
1. (a) 0.5592 (b) 0.9518 (c) 0.9886 (d) 0.4540 (e) 0.9793
   (f) 0.4438 (g) 1.0355 (h) 0.5117 (i) 7.3639
2. (a) 8.998° (b) 65.99° (c) 81.11° (d) 53.32° (e) 78.93
   (f) 47.69° (g) 2.87° (h) 15.66°
3. (a) 24.01° (b) 70° (c) 14.69° (d) 64.6°
   (e) 16.54° (f) 89.52° (g) 74.86 (h) 61.97°
4. (a) 24° (b) 73° (c) 4.4° (d) 21.69°
   (e) 19.64° (f) 62.55° (g) 0.93° (h) 39.57°

Exercise 6:4
1. (a) 0.5592 (b) 0.9527 (c) 0.9886 (d) 0.4540 (e) 0.5793
   (f) 0.4428 (g) 1.0355 (h) 1.0392 (i) 7.3639
2. (a) 9° (b) 66° (c) 81°7' (d) 53°19' (e) 78°55'
   (f) 47°43' (g) 2°52' (h) 15°40'
3. (a) 24° (b) 70° (c) 14°42' (d) 64°36' (e) 16°32' (f) 89°29'
   (g) 74°52' (h) 61°56'
4. (a) 24° (b) 73° (c) 4°24' (d) 21°41' (e) 19°37'
   (f) 62°52' (g) 0°56' (h) 39°34'

Exercise 6:5
1. $\sin \alpha = 0.36$ and $\tan \alpha = 0.5624$
2. $\sin x = \dfrac{12}{13}$, $\tan x = \dfrac{5}{12}$ 3. $\sin A = \dfrac{4}{5}$, $\cos A = \dfrac{3}{5}$

Exercise 6:6
1. (i) 18° (ii) 65° (iii) 46.5° (iv) 90° (v) 67.5° (vi) 90° 2. 0.5736
3. (a) 38° (b) 67° (c) 50° (d) 20°
4. (a) 30° (b) 56° (c) 45° (d) 0° (e) 90° 5. 0.7431 6. 0.4540
7. Any two positive numbers x and y such that $x + y = 90°$

Exercise 6:7
1. (a) $\dfrac{5}{4}$ (b) 0 (c) $\dfrac{3}{2}$ (d) $\dfrac{4}{3}$ (e) 1
2. (a) 1 (b) $2\sqrt{3}$

Exercise 6:8
1. 8 m 2. 3 m 3. 11 m 4. (a) 6 m (b) 120° (c) 10.4 m.
7. (a) 1124 m (b) 2782 m

Exercise 7:1
1. 10 cm 2. 462 cm 3. 352 cm 4. 4 cm 5. 528 cm^3
6(a) 880 cm^2 (b) 1496 cm^3 7. 562 cm^3 8. 462 cm^3 9. 314.3 cm^3
10. (a) 528 cm^2 (b) 712 cm^3 11. 45° 12. 0.007 m

Exercise 7:2
1. (a) 5544 cm^2, 38808 cm^3 (b) 1386 cm^2, 4850 cm^3
2. 1400 cm^2 3. 1:343 4. 1767 cm^3 5. 10,000
6. (a) 56.6 cm^2 (b) 56.5 cm^2 (c) 84.9 cm^2

Exercise 7:3
1. (a) 122.6 cm^3 (b) 106 cm^2 2. (a) 1685.1 cm^3 (b) 594 cm^2 (c) 7.5 cm
3. (a) $\frac{370}{3}\pi$ cm^3 (b) $12\pi\sqrt{101}$ cm^2 4. (a) 8.2 cm^3 (b) 34.9 cm^2
5. (a) 3π cm^3 (b) $\left(\frac{7}{2}\pi + 24\right)$ cm^2 6. (i) 5.9 cm (ii) 10.2 cm
7. (a) 234.7 cm^3 (b) 191.2 cm^2 8. 5011.3 cm^3 9.(a) 16..1 cm^3 (b) 47.08 cm^2
10. (a) 8382.5 cm^3 (b) 3116.1 cm^3 (c) 2464 cm^2

Exercise 8:1
1. (pie chart: Football 75, Handball, Basketball 75)

2. (a) 600 FCFA (b) 75 FCFA.

Answers to Structural Exercises

3.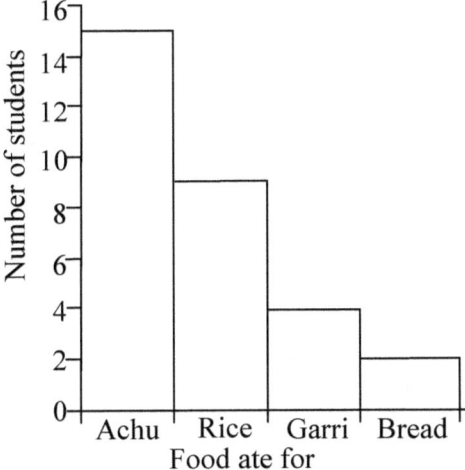

4. 41.7 %

5.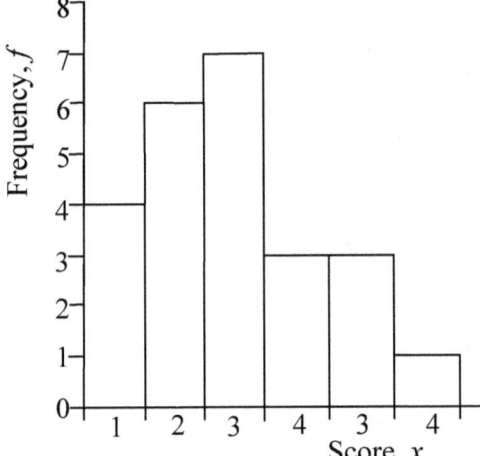

6. 14°   7. 56

Competency Based Mathematics for Secondary Schools. Book 3

8.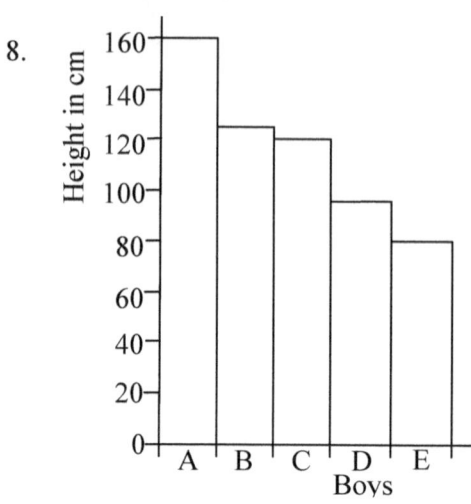

9.(a) 100° (b) 15   10. $w = 156°$, $x = 72$, $y = 16$, $z = 24°$ 11. 105°

Exercise 8:2
1. 13   2. 1   3. 9   4. (a) 54 kg (b) 51.2 kg (c) 54 kg
5. (a) 2.92 (b) 3   (c) 3   6. (a) 5.3   (b) 5   (c) 5
7. (a) 7   (b) 7.1 (c) 7   8. (a) 113.4 FRS (b) 100 FRS (c) 50 FRS
9. (a) 70 kg   (b) 68 kg   (c) 70.25 kg   10. (a) 30   (b) 8   (c) 6.1
11. (a) 2   (b) 2   (c) 2   12. (a) 5   (b) 12   (c) 7.7

Exercise 8:3
1. Mode. Stock goods are most needed.
2. Mean. It takes into account all the values.
3. Mean. It takes into account all the values.
4. The average is a misleading statistic because so many farmers have no pigs and one farmer has so many.

Exercise 8:4
1. (a)

x	66	67	68	69	70	71
f	2	3	7	4	3	1

(b) 68   (c) 68   (d) 68   (e) 68.3

2. (a)

GRADE	A	B	C	D	E	F
Angle of Sector	30	60	90	40	70	70

Answers to Structural Exercises

(b)
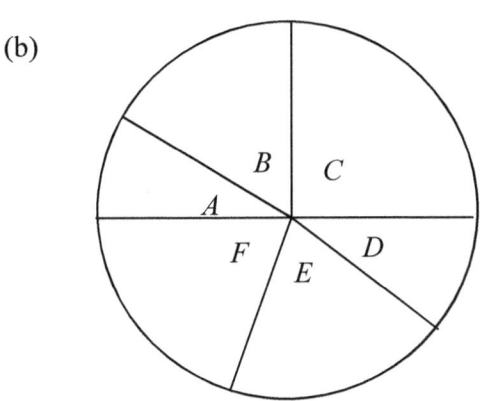

(c) (i) 72   (ii) 8   (d) 1:1

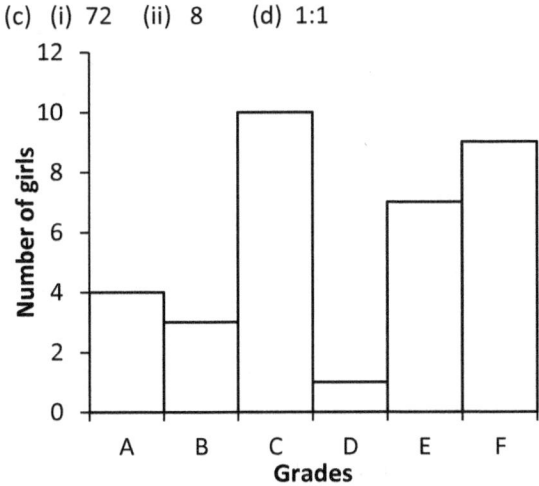

3. (a)

x	4	5	6	7	8	9	10
f	9	3	7	4	5	1	1

(b) 4   (c) 6

Exercise 9:1
1. (a) {blue, red, green, white, black}   (b) {white, green}
2. {(H,H),(H,T),(T,H),(T,T)}   3. {1,2,3,4,5,6}
4. (a) {2,3,5,7} (b) {2,4,6,8,10} (c) {1,3,5,7,9} (d) {1,4,9} (e) {3,6,9}
5. {a, e, a, i}   6. 12

Competency Based Mathematics for Secondary Schools. Book 3

Exercise 9:2

1. (a) $\dfrac{1}{6}$ (b) $\dfrac{1}{2}$ (c) $\dfrac{1}{2}$ (d) $\dfrac{2}{3}$ (e) $\dfrac{1}{2}$

2. (a) $\dfrac{3}{8}$ (b) $\dfrac{1}{8}$ (c) 0 (d) $\dfrac{1}{8}$

3. $\dfrac{1}{4}$ 4. $\dfrac{3}{10}$ 5. $\dfrac{2}{11}$ 6. $\dfrac{1}{3}$

Exercise 10:1

(a) 1. $3x+15$ 2. $4y+8$ 3. $4x+xy$ 4. $3v+uv$ 5. $2x-2$
6. $5y-15$ 7. $6p-3pq$ 8. $4t-st$ 9. $10x-2x^2$ 10. x^2+4x+3
11. x^2+x-2 12. $pq-p+3q-3$ 13. x^2+6x+9
14. x^2-4x+4 15. $16+8a+a^2$ 16. $9-6y+y^2$

(b) 1. 2401   2. 10201   3. 39204   4. 195364

Exercise 10:2

(1) $3(x+y)$ (2) $2(p+3q)$ (3) $5(x-2y)$ (4) $4y(x+2)$ (5) $3u(2v-f)$
(6) $\dfrac{1}{4}x(y+p)$ (7) $\dfrac{1}{2}x(a-b)$ (8) $\dfrac{1}{3}(2y-x)$

Exercise 10:3

(a) $(y+1)(x+1)$ (b) $(a+3)(x+1)$ (c) $(2p+1)(3x+2)$ (d) $(3x+2)(1-2y)$
(e) $(3x-2)(2y-3)$ (f) $(u-v)(5-t)$ (g) $(a-3)(m-n)$ (h) $(r+3s)(p-2q)$

Exercise 10:4

1. (a) $(x+y)(x-y)$ (b) $(1+3y)(1-3y)$ (c) $3(3)(2+x)(2-x)$
   (d) $(2a+1)(2a-1)$ (e) $(2y-3x)(2y+3x)$ (f) $(2u-5y)(2u+5y)$
   (g) $(6a-7b)(6a+7b)$ (h) $(x-yz)(x+yz)$

(2) (a) 81   (b) 976   (c) 115000   (d) 0.75   (e) 135000   (f) 0.000005

Exercise 10:5

1. $(x+3)(x-2)$ 2. $(p+1)(p+11)$ 3. $(y-4)(y-3)$
4. $(x+8)(x-2)$ 5. $(x-5)(x+3)$ 6. $(5x-3)(2x+1)$
7. $(4a+5)(a-2)$ 8. $(5-6y)(1-2y)$ 9. $(3+2x)(1-x)$
10. $(5-3x)(2-x)$

Answers to Structural Exercises

Exercise 11:1

1. $x=1, y=2$ 2. $x=1, y=3$ 3. $a=2, b=\frac{3}{2}$ 4. $x=3, y=1$
5. $x=4, y=1$ 6. $x=\frac{47}{37}, y=\frac{25}{37}$ 7. $x=0, y=2$ 8. $x=3, y=12$

Exercise 11:2

See Exercise 11:1 above

Exercise 11:3

1. (a) $p=4, q=-2$ (b) $s=7, t=4$ (c) $x=4, y=3$
   (d) $m=\frac{1}{2}, n=\frac{1}{4}$ (e) $x=3, y=1$ (f) $u=5, v=2$

2. See 1. Above

3. (a) $a=17, b=-10$ (b) $x=2, y=-1$ (c) $x=-2, y=\frac{3}{2}$
   (d) $x=1, y=2$ (e) $x=6, y=1$ (f) $x=-4, y=-2$
   (g) $x=1, y=-3$ (h) $x=4, y=3$ (i) $x=1, =3$
   (j) $a=3, b=-1$ (k) $x=5, y=2$ (l) $x=5, y=2$
   (m) $a=2, b=-3$ (n) $x=8, y=3$ (o) $a=-1, b=-1$ (p) $r=2, s=3$

Exercise 11:4

1. $x=9, y=1$ 2. $x=-\frac{1}{2}, y=1$ 3. $a=500, b=1000$ 4. $x=1400, y=450$
5. $m=700, n=200$ 6. $x=800, y=500$
7. $p=700, q=600$ 8. $u=4000, v=7000$

Exercise 11:5

1. Book = 100 Frs., pencil = 20 Frs. 2. $m=5, n=-6$ 3. $27, -1$ 4. $16, 10$
5. bottle $=17g$, cork $=1g$ 6. (a) 16g (b) 6g
7. Coconut = 130 Frs., orange = 20 Frs. 8. Man = 36 years, son = 4 years.
9. (a) Number of first-class rooms = 25,   Number of second-class rooms = 7
   (b) For first-class rooms = 112500 FRS,   For second-class rooms = 17500 FRS
10. 16 shirts and 16 trousers   11. 4 km

Exercise 11:6

1. $x=4$ or $x=-2$ 2. $x=-2$ or $x=1$ 3. $x=2$ or $x=3$ 4. $x=-\frac{1}{2}$ or $x=\frac{3}{2}$

5. $x = 7$ or $x = -3$ 6. $x = -\dfrac{2}{3}$ or $x = \dfrac{3}{2}$ 7. $x = \dfrac{5}{6}$ or $x = -\dfrac{3}{2}$

8. $x = -\dfrac{1}{2}$ or $x = 5$

Exercise 11:7
1. $x = 3$ or $x = -15$ 2. $u = -19$ or $u = -1$ 3. $p = -8$ or $p = 1$
4. $y = 13$ or $y = 5$ 5. $a = 4$ or $a = 7$ 6. $x = 33$ or $x = -3$
7. $x = -16$ or $x = 2$ 8. $x = -19$ or $x = 1$

Exercise 11:8
1. -3 and -2 or 3 and 2 2. -3 and $-3^2 = 9$ or 4 and $4^2 = 16$
3. 15, 135 FCFA   4. 15 m by 16.5 m   5. $n = -13$ or $n = 12$ 6. 3 and 9
7. 16   8. 12 m by 2 m   9. $h = 8$ m, $b = 12$ m   10. 8   11. 11 and 3
12. 12 rows

Exercise 12:1
1. (a) True (b) True (c) True (d) True (e) True (f) True (g) False (h) False
   (i) False (j) True   (k) False
2. (a) {7} (b) {8,9} (c) {2,4,6,8} (d) {5} (e) ∅ (f) {8} (g) ∅   (h) ∅ (i) {5}
   (j) {1,2,3,4,6,7,8,9}(k) {7,8,9} (l) {5,6,7,8,9}
3. (b) false, (c) false, (d) false, (g) false (h) true
4. (a) closed (b) closed   (c) open   (d) closed   (e) closed   (f) closed
   (g) closed (h) open   (i) closed   (j) open   (k) open
5. (a) {6,7,8,9} (b) {1,2,5} (c) {1,3,5,7,9}   (d) {2,4,6,8,10} (e) {2,3,5,7} (f) {3,6,7}
6. (a) T (b) T (c) F (d) F (e) T (f) F (g) T

Exercise 12:2
1.(a) It is not true that Mr. Fonche died two years ago.
   (c) They are not lazy.
   (d) Not all Bamenda people eat Achu.
   (e) Loh can drive.
   (f) Nigeria is not an African Country
   (g) History is not a science subject.
   (h) He was not the president of Cameroon.
   (i) It is not true that everyone loves Mr. Paul Biya.
   (j) She does not come from Nkambe.
   (k) It is not true that Science has done more harm than good.

Answers to Structural Exercises

2. (a) $4y \not> 12$ (b) $3+5 \neq 9$ (c) $3p-1 \not\geq 17$ (d) $4 \times 3 \neq 12$ (e) $6x+1 \not< 19$
   (f) $2x-1 \neq 0$ (g) $A \subset B$ (h) $A \cap B = \emptyset$

Exercise 12:3

1. (i) $\sim p$ (ii) $q \wedge p$ (iii) $\sim p \wedge \sim q$ (iv) $p \vee q$ (v) $\sim (p \wedge q)$ (vi) $\sim (p \vee q)$

2. (a) Mrs. Ngwa visited me.
       Mrs. Tayong visited me.
   (b) Nfor likes rice.
       Nfor likes beans.
   (c) Mr. Nkwain is a Cameroonian.
       Mr. Nkwain is an ambassador.
   (d) Bamenda is a big city.
       Bafoussam is a big city.

3. (i) Nfor is not hungry.
   (ii) Nfor is hungry and thirsty.
   (iii) Nfor is hungry or thirsty.
   (iv) If Nfor is hungry then he is thirsty.
   (v) Nfor is hungry if and only if he is thirsty.
   (vi) If Nfor is hungry then he is not thirsty.
   (vii) Nfor is neither hungry nor thirsty.
   (viii) Nfor is hungry if and only if he is not thirsty.

4. (a) If Fombe is rich then he is happy.
   (b) If he was drunk then he drank alcohol.
   (c) If it is night in Cameroon then places are dark.
   (d) If she performed well then she had a prize.

6. (a) Fombe is rich if and only if he is happy.
   (b) He was drunk is a necessary and sufficient condition that he drank alcohol.
   (c) It is night in Cameroon if and only if places are dark.
   (d) She performed well if and only if she had a prize.

6.

p	q	$p \rightarrow q$	$q \rightarrow p$	$(p \rightarrow q) \wedge (q \rightarrow p)$
T	T	T	T	T
T	F	F	T	F
F	T	T	F	F
F	F	T	T	T

p	q	p →q	q →p	p ↔q
T	T	T	T	T
T	F	F	F	F
F	T	T	F	F
F	F	F	T	T

Therefore, (p →q) ∧ (q →p) ≡ p ↔q is false.

7.

p	q	p	p	q	~p	~p ∨ q
T	T	T	T	T	F	T
T	F	F	T	F	F	T
F	T	T	F	T	T	T
F	F	F	F	F	T	F

p	q	~p	~p ∨ q
T	T	F	T
T	F	F	T
F	T	T	T
F	F	T	F

Therefore, p →q ≢ ~p ∨ q.

8. (a)

p	q	r	p ∧ q	q ∧ r	(p ∧ q) ∧ r	p ∧ (q ∧ r)
T	T	T	T	T	T	T
T	T	F	T	F	T	T
T	F	T	F	F	F	F
T	F	F	F	F	F	F
F	T	T	F	T	F	F
F	T	F	F	F	F	F
F	F	T	F	F	F	F
F	F	F	F	F	F	F

Therfore, (p ∧ q) ∧ r ≡ p ∧ (q ∧ r)

Answers to Structural Exercises

(b)

p	q	r	$q \wedge r$	$p \vee q$	$p \vee r$	$p \vee (q \wedge r)$	$(p \vee q) \wedge (p \vee r)$
T	T	T	T	T	T	T	T
T	T	F	F	T	T	T	T
T	F	T	F	T	T	T	T
T	F	F	F	T	T	T	T
F	T	T	T	T	T	T	T
F	T	F	F	T	F	F	F
F	F	T	F	F	T	F	F
F	F	F	F	F	F	F	F

Therefore, $p \vee (q \wedge r) \equiv (p \vee q) \wedge (p \vee r)$

9. (a)

p	q	$p \vee q$	$p \wedge q$	$(p \vee q) \rightarrow (p \wedge r)$
T	T	T	T	T
T	F	T	F	F
F	T	T	F	F
F	F	F	F	T

(b)

p	q	$p \vee q$	$\sim p$	$(p \vee q) \wedge \sim p$
T	T	T	F	F
T	F	T	F	F
F	T	T	T	T
F	F	F	T	F

(c)

p	q	$\sim q$	$p \wedge \sim q$	$p \rightarrow (p \wedge \sim q)$
T	T	F	F	F
T	F	T	T	T
F	T	F	F	T
F	F	T	F	T

(d)

p	q	$p \rightarrow q$	$(p \rightarrow q) \wedge q$
T	T	T	T
T	F	F	F
F	T	T	T
F	F	T	F

(e)

p	q	p→q	~p	(p→q)→~p
T	T	T	F	F
T	F	F	F	T
F	T	T	T	T
F	F	T	T	T

(f)

p	q	p→q	p∧(p→q)
T	T	T	T
T	F	F	F
F	T	T	F
F	F	T	F

A	B	A ∨ B
T	T	T
T	F	T
F	T	T
F	F	F

S	P	S∧P
T	T	T
T	F	F
F	T	F
F	F	F

Exercise 12:4

1. $\exists! x: x = 2, x \in E, p(x)$
2. $\forall x \in D, p(x)$
3. $\exists v \in F, p(x)$
4. $\nexists x \in S, p(x)$
5. $\exists! x: x = 0, x \in \mathbb{N}, p(x)$
6. $\forall s \in F, p(x)$
7. $\exists p \in R, s(x)$
8. $\nexists x \in \mathbb{R}, p(x)$

Exercise 12:5

1. r: Ngoh is a liar.
2. r: Some polygon are rectangles.
3. r: x is not in the range $4 \leq x \leq 7$ or $x \notin 4 \leq x \leq 7$
4. r: Nfor is never happy.
5. r: Bamenda is in Europe.
6. r: 18 is a multiple of 12.
7. r: Each angle in the quadrilateral ABCD is a right angle.
8. r: ABCD does not have equal sides.
9. r: Bih is not beautiful.
10. r: A = B.

Answers to Structural Exercises

Exercise 13:1

1. $w = \dfrac{bTu}{a-bT}$ 2. $-40°, C = \dfrac{5F-160}{9}$ 3. $W = \dfrac{10R}{T-5S}, 10$

4.(a) 1.5×10^{-5} (b) $h = \dfrac{cV}{\pi r^2}$ 5. $\dfrac{x}{y} = \dfrac{1}{5}$ 6. $p = 4m - 2n = 2(2m-n)$

7. (i) $g = \dfrac{hkL}{kL-h}$ (ii) $k = 2$

Exercise 13:2

1. $A = \left(\dfrac{\pi n}{V}\right)^2 + 3$ 2. $t = \dfrac{p^2 r}{(a-p)^2}$ 3. $p = \dfrac{4n^2\pi^2(k+1)}{m^2}$, $p = 64$

4. $q = \dfrac{p^2(an+b)}{n}$ 5. (a) $v = \left(\dfrac{k}{y}\right)^n$ (b) $v = \pm 2$ 6. $l = 4\pi^2 T^2 g$ 7. $a = \dfrac{v^2 s + t}{r}$

8. $m = \dfrac{n}{p^2-1}$ 9. $h = \dfrac{3V - \pi r^3}{\pi r^2}$

Exercise 13:3

1. (a) $r = \dfrac{-\pi l \pm \sqrt{\pi^2 l^2 - 4A\pi}}{2\pi}$ (b) $t = \dfrac{-u \pm \sqrt{u^2\, 2as}}{a}$

   (c) $x = \dfrac{-b \pm \sqrt{b^2 - 4ac}}{2a}$ (d) $t = \dfrac{-\pi r \pm \sqrt{A\pi + \pi r^2}}{\pi}$

2. $\dfrac{r}{s} = 1$ or $\dfrac{r}{s} = 3$ 3. $G = -\dfrac{1}{4}T$ or $G = \dfrac{T}{3}$

4. $r = \dfrac{6p}{q}$ or $r = -\dfrac{p}{3q}$ 5. $n = 2m$ or $n = \dfrac{6}{5}m$

Exercise 14:1

1. $c = 7n$ 2. $m = nk$ 3. $x = \dfrac{1}{5}y$ 4. $x = 1$ 5. $\dfrac{32}{125}$ 6. $y = \pm 12$ 7. $s = 27$

8. $p = \dfrac{3}{8}q, p = 6$ 9.(a) $k = 6$ (b) $x = \pm 2$ 10. $\dfrac{3}{4}$

Exercise 14:2

1. (a) $y = \dfrac{2}{x^2}$ (b) $\dfrac{1}{6}$ 2. $a = \pm 6, b = 9$ 3. $a = 3, b = 12$

4. $y \neq 3$, $y = 2.25$ 5. (a) 2 (b) $y = 256$ 6. $x = \pm \dfrac{2}{3}$

7. $a = 275$, $b = \dfrac{5}{6}$ 8. $y \neq 2$, $y = \dfrac{16}{9}$

9.
x	0	5	10	15	20	25
y	0	1	2	3	4	5

Exercise 14:3

1. $l = \dfrac{3}{2} mn$ 2. (a) $A = 6\dfrac{B}{C}$ (b) $A = 5$

3. (a) $y = 21xz$ (b) $x = \dfrac{5}{189}$ 4. $x = \dfrac{5}{2}$

Exercise 15:1

1. (a) $D = \{0, 1, 2, 3\}$, $R = \{0, 3, 6, 9\}$
   (b) $D = \{0, 1, 4\}$, $R = \{-2, -1, 0, 1, 2\}$
   (c) $D = \{1, 2, 3\}$, $R = \{15, 20, 25\}$
2. (a) $\{(1,2),(2,3),(3,4),(4,5),(5,6)\}$
   (b) $\{(1,3),(2,2),(3,1),(4,0),(5,-1)\}$
   (c) $\{(1,-1),(2,-2),(3,0),(4,2),(5,4)\}$
   (d) $\{(1,2),(2,5),(3,10),(4,17),(5,26)\}$
3. (a) $\{(-4,4),(-2,2),(0,0),(2,-2),(4,-4)\}$
   (b) $\{(-4,-8),(-2,4),(0,0),(2,4),(4,8)\}$
   (c) $\{(-4,2),(-2,0),(0,-2),(2,0),(4,2)\}$
   (d) $\{(-4,-3),(-2,1),(0,1),(2,3),(4,5)\}$

4.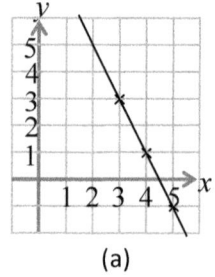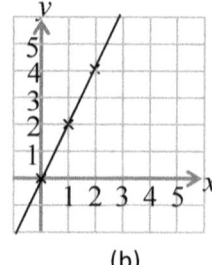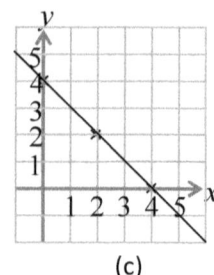

(a) (b) (c)

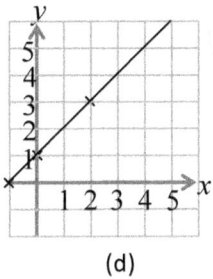
(d)

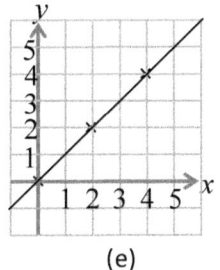
(e)

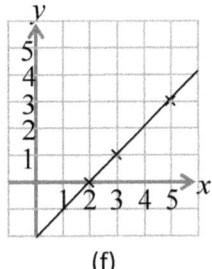
(f)

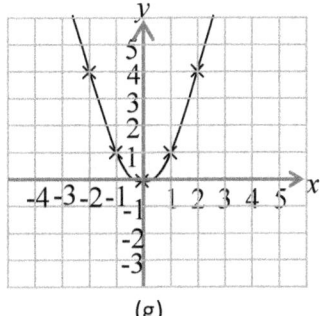
(g)

5.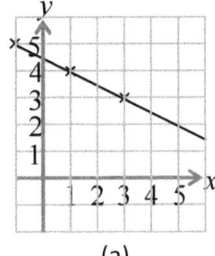
(a)

(b)

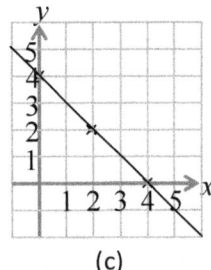
(c)

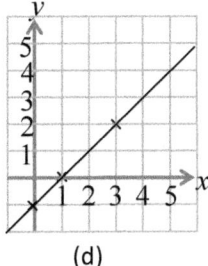
(d)

(e)

(f)

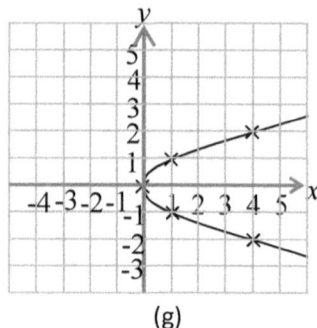

(g)

6. {(1,2),(1,4),(1,6),(3,2),(3,4),(3,6),(5,2),(5,4),(5,6)}
7. (a) {(−2,10),(−2,20),(1,10),(1,20),(4,10),(4,20)}
   (b) {(10,−2),(10,1),(10,4),(20,−2),(20,1),(20,4)}
   (c) {(−2−2),(−2,1),(−2,4),(1,−2),(1,1),(1,4),(4,−2),(4,1),(4,4)}
   (d) {(10,10),(10,20),(20,10),(20,20)}

8.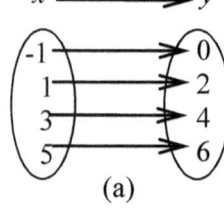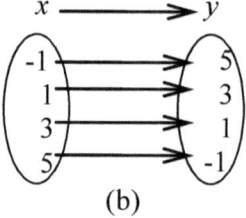

   (a)   (b)

   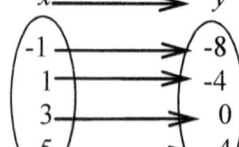   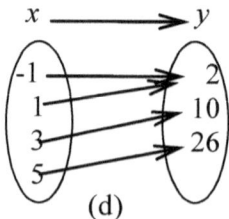

   (c)   (d)

9. (a) 'is a husband of'   (b) 'is a factor of'   (c) 'is a student of'
   (d) 'is one fifth of'   (e) 'is two less than thrice'
10. (a) Any number is a factor of itself.
    (b) $u = 4, v = 36, w = 12, x = 16$

Answers to Structural Exercises

Exercise 15:2

1. (i)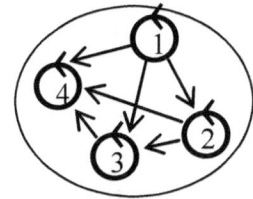

    (ii) $\forall x, y \in A, x\Re y \not\Leftrightarrow y\Re x$ i.e. the relation is not symmetric.

2. (i)

Reflexive	(a), (b), (c), (d), (f), (g)
Symmetric	(e), (f), (g), (h)
Transitive	(a), (b), (c), (d), (f), (g)
Anti-symmetric	(a), (b), (c), (d)

   (ii) (f) and (g)

Exercise 15:3

1. (a), (b), (f), (g), (h), (i)

2. (a)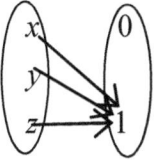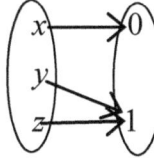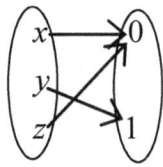

   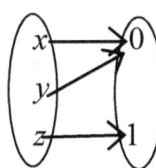 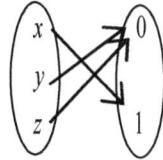

   (b) 8

3. (i) (a)

(b) 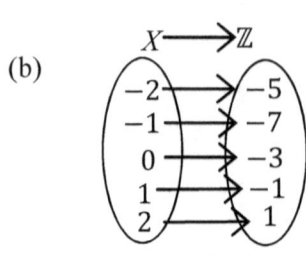　　(c)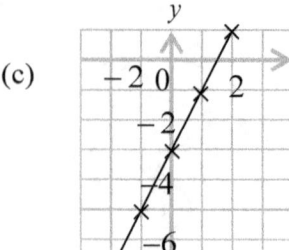

(ii) Domain = X, Range = {−7,−5,−3,−1,1}

Exercise 15:4
1. (a) (ii)　　(b) (i), (iii)　　(c) none　　(d) all
2. (i) many-one　(ii) one-one　(iii) many-one　(iv) one-one
3. (i) none　　(ii) all
4. (a) (i) range = {2,3,5}　(ii) f is an 'into' but not an 'onto'
5. injections: None;　Surjections: (a), (b), (f), (g), (h);　Bijections: (i)
6. (a) Not a function　　(b) Function, surjective　(c) Function, surjective
   (d) Function, surjective　(e) Not a function　(f) Function, surjective

Exercise 15:5
1. (a) a　(b) $R = \{2,3,5\}$　2. (i) $f^{-1}: x \mapsto \frac{x-1}{2}$　(ii) $\frac{4}{3}$　3. 0 or $\frac{2}{3}$
4. (a) $gf: x \mapsto \frac{6x-1}{3}$　(b) $g^{-1}: x \mapsto \frac{3x-1}{2}$
5. (i) 1　(ii) ±3　6. (i) 2　(ii) $x = -1$ or -2
7. (a) $\{10, 0, -2\}$　(b) $D = \{3, 0, 1\}$　8. (a) -5　(b) ±2
9. (a) 1　(b) $f \circ g(x) = x^2 + 4x + 1$　(c) $g^{-1}: x \mapsto x - 2$　(d) $p = 3, q = -1$
10. (a) $f^{-1}: x \mapsto x - 2$　(b) $g(\sqrt{2}) = 1, hf(0) = 28$　(c) $x = 0$ or $x = \frac{2}{7}$
11. (i) $gf(x) = 3x^2 - 18$　(ii) $9x^2 - 36x + 50 = 0$　(iii) $x = 0$ or $x = 4$
12. (i) Domain = $\mathbb{R} - \{2\}$, Range = $\mathbb{R} - \{-1\}$　(ii) $g^{-1}: x \mapsto \frac{x+1}{2}$　(iii) $h: x \mapsto \frac{x+2}{x-1}$
13. (a) $pr(x) = 4x^2 + 1$　(b) $2x^2 + 2$　(c) $x = 1$ or $x = 2$　(d) $x \neq 1$
    (e) $p^{-1}: x \mapsto \pm\sqrt{x-1}$,　$q^{-1}: x \mapsto 3(x-1)$
14. (a) -1　(b) $f^{-1}: x \mapsto 1 - x$　(c) $\frac{1 \pm \sqrt{5}}{2}$
15. (a) $\mathbb{R} - \{-3, 3\}$　(b) $\{x: x \leq -2 \text{ or } x \geq 2, x \in \mathbb{R}\}$　(c) $\mathbb{R}$　(d) $\mathbb{R} - \{0\}$
    (e) $\mathbb{R} - \{-3, 2\}$　(f) $\mathbb{R}$

www.ingramcontent.com/pod-product-compliance
Lightning Source LLC
Chambersburg PA
CBHW071410180526
45170CB00001B/50